METHODS IN AGRICULTURAL CHEMICAL ANALYSIS

ANALYSIS

A Practical Handbook

To Alwyn, Antonia and Bethan

METHODS IN AGRICULTURAL CHEMICAL ANALYSIS
A Practical Handbook

N.T. Faithfull

Institute of Rural Studies
University of Wales
Aberystwyth
UK

CABI *Publishing*

CABI *Publishing* is a division of CAB *International*

CABI Publishing
CAB International
Wallingford
Oxon OX10 8DE
UK

CABI Publishing
10 E 40th Street
Suite 3203
New York, NY 10016
USA

Tel: +44 (0)1491 832111
Fax: +44 (0)1491 833508
E-mail: cabi@cabi.org
Web site: www.cabi-publishing.org

Tel: +1 212 481 7018
Fax: +1 212 686 7993
E-mail: cabi-nao@cabi.org

A catalogue record for this book is available from the British Library, London, UK.

Library of Congress Cataloging-in-Publication Data
Faithfull, N.T. (Nigel T.)
  Methods in agricultural chemical analysis: a practical handbook / N.T. Faithfull.
      p. cm.
Includes bibliographical references (p. 206).
  ISBN 0–85199–608–6
  1. Soils--Analysis--Handbooks, manuals, etc. 2. Plants--Analysis--Hand-books, manuals, etc. 3. Chemistry, Analytic--Handbooks, manuals, etc. I. Title.
S593 .F19 2002
630′.2′43--dc21
2002005768

ISBN 0 85199 608 6

Typeset by Wyvern 21 Ltd, Bristol
Printed and bound in the UK by Biddles Ltd, Guildford and King's Lynn

Contents

Preface

The need for this publication has arisen in four ways. The first is that relatively few staff engaged in agricultural research in educational institutions have sufficient knowledge of chemistry to make informed decisions regarding choice of the most suitable analytical method for their purposes. For example, an unsuitable sample drying process can destroy or seriously degrade the component being estimated. Second, there has been a need for a book containing methods of soil and crop analysis suitable for use in undergraduate practical classes. Lecturers under pressure to carry out publishable research and burdened with administrative duties have little time for scouring libraries and the Web for such methods. For the benefit of those lacking much experience in laboratory experimentation, the methods are described in greater practical detail than found in many publications. Third, the useful manual *The Analysis of Agricultural Materials*, MAFF/ADAS Reference Book 427, HMSO, 1986, is now out of print. Lastly, the growth in organic farming, and the establishment of the Organic Farming Centre for Wales, funded by the National Assembly for Wales and based in the University of Wales, Institute of Rural Studies, Aberystwyth, has engendered a fresh interest in analytical methods more suitable for sustainable agriculture, and a chapter is included on this area of analysis.

The nature of the contents will be determined by the practicability of the methods in undergraduate teaching, by their acceptability for research publications, and by their affordability by public sector institutions. The use of very expensive instruments may be referred to, but not described in detail. This background knowledge will assist the choice of whether to send samples away for analysis. The methods have not been chosen for their suitability in legal proceedings, although references to such will be made, and where published on the World Wide Web, the respective websites will be given. These official methods tend to be more elaborate, longer to perform, and far more rigorous than required in our case. The use of the web is growing apace, and website addresses will often be inserted in the text to aid further research. There is no attempt to include every possible procedure, but to provide the most useful selection.

It is anticipated that another author will publish a volume concentrating on chemical analysis dealing with ruminant animal nutrition. To avoid duplication, this volume will not cover that area in depth.

Soils and composts

Analyses will be those in common use for soils from fields for both grass and arable crops. MAFF/ADAS (now DEFRA) methods and Index Tables are reproduced by permission of the Controller of Her Majesty's Stationery Office (Ref. 20001327). Analyses for nitrogen mineralization are included. Special consideration will be given to composts and recycled urban waste.

Fertilizers

It is quite common for the researcher to check the specified minerals as stated on the label. Some methods will cover the usual elements.

Plant materials

Research methods demand large numbers of samples. The only way the throughput can be achieved is by using some form of automatic processing. Such methods using segmented flow analysers were conceived by Skeggs (1957) for use in clinical analysis, but found wide application in water, soil and plant analyses in the mid-1960s. In 1968, the author established an analytical laboratory at the University of Wales, Aberystwyth, with the remit to install this type of equipment for the analysis of plant samples from the Departments of Agriculture, Biology and Biochemistry. Further discussion of segmented flow analysis will be found in Chapter 1. These are the reasons why, in addition to the well-established manual procedures such as for fibre, automatic methods will be preferred if they exist, but references to the equivalent manual methods will usually be provided if available.

Feeds

This relates closely to animal nutrition, which may in the future be published

elsewhere as indicated previously. Considerations relating to grass, hay, haylage, silage, compound feeds and grain will be given.

Plant components

Research samples may contain just one part of the structure of the plant: leaf,
stem, root, etc. Certain precautions may be applicable.

Biological substances

This will not be covered in depth, but some types of material may occasionally be presented for analysis, so a few selected procedures and references will be given.

Equine nutrition

In some ways this is a developing area, and certainly lags behind ruminant
nutrition in the published material. The discussion will include any details of
recent work in this field.

Microbiological analysis

This is really not chemical analysis, but some references are suggested.

About the Author

Nigel Faithfull spent 4 years in the laboratories of RTZ at Avonmouth before proceeding to the University of Wales at Aberystwyth. He graduated with Honours in Chemistry (1968), and immediately took charge of the newly established Agricultural Sciences Analytical Laboratory. Research into automated methods in herbage analysis led to an MSc, and further studies involving atomic absorption spectrophotometry resulted in a PhD in 1975. He is a Chartered Chemist and a Member of the Royal Society of Chemistry. He was a member of college safety committees for about 30 years. Following a merger with the Welsh Agricultural College, the Analytical Laboratory is now located at the UWA Institute of Rural Studies, Llanbadarn Campus, Aberystwyth, where a commercial soil analysis service for farmers has been in operation for several years.

N.T. Faithfull, MSc PhD (Wales) CChem MRSC
Institute of Rural Studies
University of Wales
Llanbadarn Campus
Aberystwyth
Ceredigion SY23 3AL
UK

Disclaimer

Trade names in this publication are used only to provide specific information and to illustrate the type of equipment being discussed. Mention of a trade name does not imply an endorsement of that product or constitute a recommendation of it in preference to any other product which is not mentioned. The purchaser of any equipment must ensure it is suitable for the purpose for which it is intended, and compatible with any items to which it is to be connected.

All methods should be carried out only by competent persons and with adequate supervision when necessary. All obligations under The Control of Substances Hazardous to Health Regulations 1999 (COSHH), should be observed, and risk-assessment documentation completed. Appropriate personal protective equipment should be provided and worn whenever recommended. Persons carrying out the procedures in this manual do so entirely at their own risk, and neither the author, publishers, or anyone mentioned in, or connected with this publication can be held in any way responsible for any accidents no matter how caused.

Acknowledgements

First, I would like to express my gratitude to the teachers, lecturers and industrial scientists who have instilled a high regard for practical analytical chemistry, with the need for care, accuracy, and the development of a skilful and safe technique.

There are a number of individuals who have been most helpful in my search for information. It is difficult to remember everyone I have consulted over a period of 18 months, and my sincere apologies for any omissions. They include the following, with the area of advice in parentheses:

Professor W.A. Adams (soil science), University of Wales, Aberystwyth, UK

F.M. Balzer (analysis for organic farming), Labor Dr F.M. Balzer, Wetter, Germany

Zoltán Bodor (fertilizer analysis), Kemira, Finland

Joao Coutinho (soil sulphate analysis), Universidade de Trás-os Montes e Alto Douro (UTAD), Portugal

Steve Cuttle (soil analysis for organic farming), Institute for Grassland and Environmental Research, Aberystwyth, UK

Sue Fowler (organic farming), Institute of Rural Studies, University of Wales, Aberystwyth, UK

Professor D.I. Givens (NIR), ADAS Nutritional Sciences Research Unit, Stratford-upon-Avon, UK

John Hollies (soil analysis), The Potash Development Association, Laugharne, Carmarthen, UK

D. Iorwerth Jones (cellulase digestibility)

Sue Lister (NIR), Institute for Grassland and Environmental Research, Aberystwyth

Bob Llewelyn (organic manure analysis; ADAS analytical methods), ADAS
    Laboratories, Wolverhampton, UK
Peter J. Loveland (ADAS and BCSR soil analysis approaches), Silsoe Campus,
    Cranfield University, UK
Ramadan Al-Mabruk (animal nutrition) Institute of Rural Studies, University
    of Wales, Aberystwyth, UK
Isabel McMann (references), Bodleian Library, Oxford, UK
Tim Meeks (information), Tennessee Valley Authority
Meriel Moore-Colyer (equine nutrition), Institute of Rural Studies, University
    of Wales, Aberystwyth, UK
Cornelia Moser (reference), Landwirtschaftlicher Informationsdienst (LID),
    Bern, Switzerland
Suzanne Padel (organic farming), Institute of Rural Studies, University of
    Wales, Aberystwyth, UK
Dan Powell (organic farming), Aberystwyth, UK
David Rowell (soil analysis), Department of Soil Science, University of
    Reading, UK
Mustapha Bello Salawu (digestibility spreadsheets; silage VFA analysis),
    Institute of Rural Studies, University of Wales, Aberystwyth; now at
    Commonwork Group Ltd, Kent, UK
Steve Smith (reference), Stapledon Library, Institute for Grassland and
    Environmental Research, Aberystwyth, UK
Stuart Smith (autoanalysis methodology), Divisional Manager, BL-Analytics,
    Bran+Luebbe, UK
Rebecca Stubbs (editorial support), CABI *Publishing*, Wallingford, UK
Claudine Tayot (reference), OPOCE online helpdesk
Paul Thomas (NSP methodology), Institute for Grassland and Environmental
    Research, Aberystwyth, UK
Ann Vaughan (lab techniques), Institute of Rural Studies, University of Wales,
    Aberystwyth, UK
Joanne Vessey (fertilizer analysis), Hydro Agri (UK) Limited, Immingham Dock,
    UK
Keith Way (NCGD method), Laboratory Consultant
Gabriele Weigmann-Dramé (reference), VDLUFA-Verlag, Darmstadt,
    Germany
Lorraine Whyberd (information), Royal Society of Chemistry, London, UK
David Wilman (Fig. 1.5; agronomy), Institute of Rural Studies, University of
    Wales, Aberystwyth, UK

    I appreciate the efforts of the staff of the Huw Owen and Thomas Parry
Libraries, University of Wales, Aberystwyth, for retrieving large old tomes from
the store in an attempt to trace the origins of some methods. I am also grate-
ful to the university for permission to use the facilities of the Information
Services subsequent to taking early 'retirement'.
    I would like to thank ADAS, Wolverhampton, for permission to repro-
duce the information in Appendix 7; Bran+Luebbe, for permission to reproduce the methods in Appendices 5 and 6; and the Controller of Her Majesty's

Stationery Office for a licence to reproduce material from the MAFF/ADAS publications: *The Analysis of Agricultural Materials* Reference Book 427, and *Fertiliser Recommendations* Reference book 209 5th and 6th edns.

Finally, many thanks to my wife Eileen, for her continuous support, patience and encouragement throughout this task.

Abbreviations and Acronyms

AA	atomic absorption
AAS	atomic absorption spectrophotometry
ac	alternating current
ADAS	Agricultural Development and Advisory Service (UK)
ADF	acid detergent fibre
AES	atomic emission spectrometry
AOAC	Association of Official Analytical Chemists
BCSR	basic cation saturation ratio
CEC	cation exchange capacity
cmol$_c$	centimole charge
COSHH	Control of Substances Hazardous to Health Regulations (1999)
CP	crude protein
CRM	Certified Reference Material
CSL	Central Science Laboratory
DEFRA	Department for Environment, Food and Rural Affairs (UK)
detn	determination
DM	dry matter
DMD	dry matter digestibility
DMSO	dimethylsulphoxide
DOMD	digestible organic matter in dry matter

DTPA	diethylenetriaminepentaacetic acid, or diethylenetrinitrilopentaacetic acid
ECEC	effective cation exchange capacity
ED	effective degradability
EDTA	diaminoethanetetraacetic acid, ethylenediaminetetraacetic acid, or (ethylenedinitrilo)tetraacetic acid
EPA	Environmental Protection Agency (US)
FAPAS	Food Analysis Performance Assessment Scheme
FEPAS	Food Examination Performance Assessment Scheme
FIA	flow-injection analysis
FP	flame photometry
GC	gas chromatography
GLC	gas–liquid chromatography
GLP	Good Laboratory Practice
HDPE	high density polyethylene
HPLC	high-performance liquid chromatography
HS	humic substances
HSE	The Health and Safety Executive (UK)
ICP	inductively coupled plasma emission spectroscopy
ICP-MS	hyphenated technique of ICP followed immediately by MS
ICP-OES	inductively coupled plasma optical emission spectroscopy
ID	inside diameter
IGER	Institute for Grassland and Environmental Research
IMS	industrial methylated spirits
IR	infrared
ISE	ion-selective electrode
ISO	International Organization for Standardization
IV	*in vitro*
IVDMD	*in vitro* dry matter digestibility
LGC	Laboratory of the Government Chemist
m	mass (when used in % m/m)
MADF	modified acid detergent fibre
MAFF (now DEFRA)	Ministry of Agriculture, Fisheries and Food (UK)
MBT	mobile bag technique
ME	metabolizable energy
mol	molecular weight in grams (relative molecular mass)
MS	mass spectrometry
NAMAS	National Measurement Accreditation Service (now FAPAS)

NCGD	neutral cellulase plus gamanase digestibility
NDF	neutral detergent fibre
NIR; NIRS	near infrared, or near infrared reflectance spectroscopy
NIST	National Institute of Standards and Technology
NMR	nuclear magnetic resonance
NSP	non-starch polysaccharide
NVLAP	National Voluntary Laboratory Accreditation Program
OD	outside diameter
OECD	Organisation for Economic Co-operation and Development
OEM	original equipment manufacturer
OMD	organic matter digestibility
OSHA	Occupational Safety and Health Administration (US)
PAH	polycyclic aromatic hydrocarbon(s)
PPE	personal protective equipment
PTFE	polytetrafluoroethane
PVC	polyvinyl chloride
RI	refractive index
rpm	revolutions per minute
RPRs	reactive phosphate rocks
SGS	Société Générale de Surveillance
SI	Statutory Instrument
SLAN	sufficiency level of available nutrient
SNV-D	standard normal variate and detrend
SOC	soil organic carbon
TCA	The Composting Association
TEA	triethanolamine
TEB	total exchangeable bases
TOM	total organic matter
t.p.i.	threads per inch
UKAS	United Kingdom Accreditation Service
UKASTA	United Kingdom Agricultural Supply Trade Association
URL	Uniform Resource Locator (Internet website address)
USDA	United States Department of Agriculture
UV	ultraviolet
v	volume
VAM	valid analytical measurement
VFA	volatile fatty acid
WPBS	Welsh Plant Breeding Station (now Institute for Grassland and Environmental Research)
WSC	water soluble carbohydrate

1 Experimental Planning

Experimental Design

Experimental design is not directly related to chemical analysis, but it is important in that it determines the number of samples for processing. This could mean that there are too many tests for the laboratory to fit into its schedule, bearing in mind that there are many other customers clamouring for laboratory services. It could also mean that the cost is prohibitive for the funds available for the project.

Some of the books on the design of scientific experiments appear far too theoretical for use in college field trials. However, three books in particular have proved useful in this Institute:

- *Statistical Procedures for Agricultural Research*, 2nd edn. Gomez, K.A. and Gomez, A.A. John Wiley & Sons, 1984.
- *Agricultural Experimentation*. Little, T.M. and Hills, F.J. John Wiley & Sons, 1978.
- *Statistical Methods in Agriculture and Biology*, 2nd edn. Mead, R., Curnow, R.N. and Hasted, A.M. Chapman and Hall, 1993.

For example, the book by Gomez and Gomez describes many possible designs such as the Latin square and the lattice designs. The former can handle simultaneously two known sources of variation among experimental units. Chapters deal with 'Sampling in experimental plots', and the 'Presentation of research results'.

Plot size

The field plot size is chosen to give the required degree of precision for measurement of the selected characteristic. Sampling only a fraction of the plot, providing the sampling error is acceptable, may save time and expense. The sampling error is the difference between the value of the fraction and the value if the whole plot (population) had been sampled. If adequate precision is retained, it may be possible to bulk samples together at a later stage to reduce the numbers for chemical analysis.

Equipment Considerations

Autoanalysis

There is usually no problem of access to basic laboratory instruments and associated glassware, however, the only means of handling large numbers of tests is to apply some form of automation. An added advantage is that it improves the analytical precision and reproducibility. The most suitable technique has been based on the segmented continuous-flow principle invented by Skeggs (1957), and which was first marketed as the Technicon® AutoAnalyzer. The system consists of a number of modules powered from a stabilized 110 V supply, and a typical layout is shown in Fig. 1.1.

This was improved with the next generation AutoAnalyzer II, which provided the peristaltic pump with a metering air-bar. This aided a more regular bubble pattern with further improvement in precision. The current AutoAnalyzer 3 system offers several useful features. The Compact Sampler has random access, which means that if there is an over-range sample, which may distort the succeeding two peaks, the software will automatically instruct the sampler to repeat the affected peaks. This system saves a lot of time because the operator does not have to work out the repeats after a long run and reload the cups to be repeated. The pump in the current model has the option of dilution valves that allow automatic rerun of off-scale samples at a higher dilution. The segmented stream can pass through the colorimeter flowcell without debubbling, the software switches off the detection signal when a bubble is present. The redesigned flowcell has a square-edge planar window and uses fibre optics to ensure parallel light transmission and hence a reduction of interference from variation in refractive index of the liquid stream.

More information can be found on the manufacturer's website:

http://www.bran-luebbe.de/en/autoanalyzer.html

The price range for a basic system with a colorimeter is about £20k to £27k depending on options (e.g. PC and flame photometer) and whether educational discount applies.

Other manufacturers of segmented-flow analysers are Burkard Scientific, see:

http://www.burkardscientific.co.uk/Analytical/Systems_Analysers_SF A2000.htm

(a)

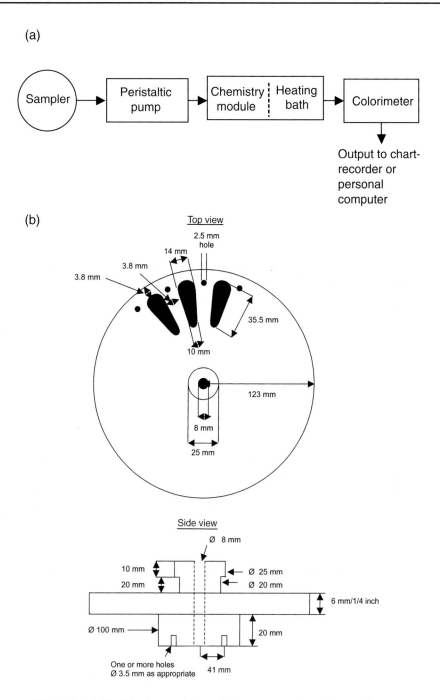

Fig. 1.1. (a) Modular layout of a typical segmented continuous flow system. (b) Simplified design for a 40-place tray to hold 8.5 ml industrial type auto-analyser cups (not to scale). It would be useful to number the cup positions. The 2.5 mm holes are for the staple which sets the stopping position of some models of sampler. ø, diameter.

and Skalar (UK Ltd), who publish a comprehensive soil and plant analysis manual, see:

http://www.skalar.com/uk/products2-1.html

A micro-bore analyser is manufactured by Astoria-Pacific Inc. USA, see:

http://www.astoria-pacific.com/analyzer.html

and is marketed in the UK by Advanced Medical Supplies Ltd, see:

http://www.ams-med.com/

It is possible to build a basic system with chart-recorder output using components from various manufacturers. Suppliers of used equipment are another possible source. Sometimes there are equipment auctions but, having learnt from bitter experience, unless one can actually go and see (and preferably test) the items listed in the catalogue, this method of purchase should be avoided. Very often parts will be missing and, being obsolete, no longer obtainable. Used equipment suppliers always include some form of guarantee, and that is worth its cost. Some used or refurbished equipment suppliers are listed in Appendix 1.

Samplers

In addition to the above manufacturers, autosamplers (Series 4000) may be obtained from:

Hook & Tucker zenyx, Harwood House, Clarendon Court, Carrs Road, Cheadle, Cheshire SK8 2LA, UK

Tel: +44 (0) 161 428 0009 Fax: +44 (0) 161 428 0019 (Price range is about £2.8k to £3.0k).

Peristaltic pumps

In addition to the above manufacturers, suitable peristaltic pumps with a minimum of 12 tube channels, such as the Ismatec IPC-16 and IPC-24 versions (cost £2000+), are obtainable from:

Bennett Scientific: http://www.bennett-scientific.com/ismatec/peri.htm and from Cole Parmer Instrument Co. Ltd at: http://www.coleparmer.co.uk/

Other multi-channel pumps are manufactured by Watson-Marlow Bredel (the 200 Series)

http://www.watson-marlow.com/wmb-gb/index.htm and distributed by Fisher:

http://www.catalogue.fisher.co.uk/

where one can browse the catalogue without needing to complete the registration form; or from Patterson Scientific:

http://www.patterson-scientific.co.uk/index.htm

Another brand is the Cole Parmer Masterflex, which can accept a maximum of 12 cartridges to give 12 channels.

The peristaltic pump is fitted with colour-coded PVC tubes of varying diameters. The flow rate is governed by the diameter and indicated by the colours of the collar at each end. Standard quality is usually of acceptable tolerance, but flow-rated tubing with a higher precision is available. In addition to PVC, other tube materials are available. Silicone rubber is free of additives and plasticizers and less likely to age over time. Solvent resistant yellow PVC retains its flexibility when used to pump solvents, and vulcanized black rubber tubing is used with concentrated acids.

Chemistry module

The outflow end of the pump tubing is connected directly to the components of the chemistry module. This module consists mainly of connectors and glass mixing coils. Proprietary modules are available, of course, but it is perfectly feasible to assemble the necessary components on a plastic tray fitted with four legs. Pump tubing and connectors are available from many suppliers. Apart from the OEMs, sources include:

Gradko International Ltd: http://members.aol.com/gradkoin/
homepage.htm, who can also supply refurbished modules.

Industrial 8.5-ml autoanalyser cups (Part No. 127-0080-01) are available from:

Gradko (see above); or

LIP (Equipment & Services) Ltd, 111 Dockfield Road, Shipley, West Yorkshire BD17 7SJ, UK. Tel: +44 (0) 1274 593411 Fax: +44 (0) 1274 589439.

The advantage of the 8.5 ml as opposed to the 2-ml or 4-ml conical cups, is that it is easier to pour into them, several analyses are possible before they need refilling, and they are more easily washed if reuse is considered. The snag is that the 40-place 8.5-ml industrial cup sample trays are no longer made by Bran+Luebbe, but Gradko can supply them for about £112 each.

In case these sample trays become unavailable, a dimensional diagram is given in Fig. 1.1b of a simplified version. The original trays were made from a glass fibre filled resin. It is suggested that suitable materials would be Nylon 66 rod, 25 mm diameter for the handle; cast Nylon 6 rod, 100 mm diameter for the underside; cast Nylon 6 available as $10 \times 500 \times 500$ mm sheet, sufficient for four trays. These are available from RS Components Ltd, at the website: http://rswww.com

Pump tubing is supplied by the above sources, also Elkay Laboratory Products UK Ltd:

http://www.elkay-uk.co.uk/

Heating bath and dialyser

These modules are options that can be incorporated within the chemistry module with newer systems, but stand-alone units are also possible.

Colorimeter and spectrophotometer

Many types are available, single or dual channel, expanded absorption ranges, digital or analogue, linear or logarithmic output, etc. If purchasing a complete new system, then the colorimeter is to be preferred. It is designed for the job with excellent long-term stability and freedom from drift; also, the sensitivity will suit the recommended chemistry, and the signal output will be compatible with the software and hardware. If building a system from variously sourced modules, the spectrophotometer will be a more useful choice. This is because it will accept standard flowcells, can be adjusted to any wavelength within its range, and usually has an output suitable for a chart-recorder. A spectrophotometer is also likely to have a scale-expansion facility allowing the measurement of absorbance values in excess of 1.0 Å, perhaps to 2.0 Å, and enabling lower values of possibly 0.1 Å to have the sensitivity increased to give a full-scale reading. This saves a lot of extra work in diluting or concentrating sample solutions. A colorimeter requires a separate filter for each wavelength. Interference filters are often required in pairs, and can be expensive.

The one essential component is the flowcell (flow-through cell), which must either be the manufacturer's own special fitting, or else a more universal design (e.g. 12.5 mm external square cross-section) as is common with most spectrophotometers. They are available with an optional built-in debubbler. These flowcells for continuous flow analysis must not be confused with flow-cells with tube connections at the top and bottom of the cell, which merely allow filling and emptying by means of an external syringe mechanism. There will be two (or three with a debubbler) connections at the top of the cell. If the wavelength is to be in the UV region, a quartz or silica cell is required, otherwise an optical glass cell is adequate. The internal cell dimensions should be cylindrical, and a path length of 10 mm × 3 mm diameter giving a volume of 0.07 ml is usually suitable. This is a micro flow-through cell. A larger cell would cause too much internal mixing and interference between wash and samples, but a smaller (ultra-micro) cell volume would emphasize noise from differences in refractive index unless specified for low flow rate methods and the particular measuring instrument.

Some manufacturers are:
Hellma Cells: http://hellma-worldwide.com/tochter/Tochter2.htm
to get the website for your area. For UK use:
http://www.hellma.demon.co.uk/
Optiglass Ltd (Starna® Brand): http://www.optiglass.co.uk/

Chart-recorders

A complete new system would have the benefit of system-control and data processing via a personal computer and proprietary software. An in-house system, however, would probably output readings to a chart-recorder. This can give a further opportunity of adjusting the scale-expansion to accommodate extra-low or extra-high peaks. It is useful to be able to vary the chart speed; this will allow the peak width to be kept at an optimum width despite any variation of the sampling rate with different methodologies. Continuous-flow methodologies mean that the recorder is left running unattended for long periods. It is vital that the sprocket pins at each end of the chart paper drive are long enough to engage positively in the holes in the paper. It is annoying to find that they slip out, perhaps at one end, and an hour's readings are wasted. It could be that the holes are fractionally out of sync with the pins, or that the paper has buckled at the end. We found with our in-house systems that a friction drive avoided these problems. The Houston Instrument OmniScribe® is of the friction type. Alternatively, a couple of large bulldog-clips attached to the end of the chart that overhangs the bench may solve the problem.

Chart reader

The reading of hundreds of peaks from a chart trace can be daunting. It is facilitated by means of a simple device known as a chart reader, apparently no longer available. It is a clear plastic A3 size sheet, originally having 15 sections, each consisting of ten vertical lines, which is laid over the chart. A baseline is first drawn on the chart under the peaks by linking the trace from aspirating the wash solution between tray changes. This compensates for baseline drift. The bottom of the vertical lines on the reader are next aligned with the baseline as it passes under the peaks from the standard solutions, which are included at the start, then after each tray to compensate for any change in sensitivity. The heights of the peaks are marked on the reader with a black grease-pencil (e.g. Royal Sovereign 808 Chinagraph) and labelled with the corresponding concentration. A connecting line is drawn to link the marks to give a standard curve. This is checked for each set of standards and corrected if necessary. The reader is laid on the chart, the bottom of the vertical lines aligned with the baseline, and the curve aligned with the top of the sample peak. The corresponding concentration is read off. A way to make a chart reader is given below.

1. Use a computer graphics program to draw eight sets of ten lines. This is printed in duplicate in portrait mode onto two sheets of laser transparency film. Corel Draw™ has a Graph Paper tool on the Polygon tool flyout. Select 40 columns and one row (the maximum number of columns is 50). Drag a rectangular graph to fill the left half of the page, make a copy and paste to the right as closely in line as possible. Go to Arrange and then Align and Distribute. Select left-hand image and align top to grid; repeat with right-hand image. Align right side of left image to grid, also left side of right image, and they should now be perfectly joined together. Select both and Group together. Adjust line width to 0.20 mm. Now draw a vertical line and adjust height to that of the graph, and line width to 0.60 mm. Copy, paste and drag to lie exactly over every tenth line. Save to file.

2. Print duplicate copies using a laser printer on to laser transparency film.
3. Guillotine a vertical edge of each copy 5 mm from the thicker border line so that the two copies will form a single graph when placed together with edges overlapping and the thicker lines aligned. Tack together with a minimum of adhesive at the top, bottom and centre.
4. Laminate using 250 μm gloss film.

Flow injection

The other type of automated wet-chemistry analysis is flow injection analysis (FIA), which was first described by Ruzicka and Hansen (1975). This is a non-segmented continuous flow method – i.e. no air bubbles are introduced to aid mixing and to separate sample and wash segments. The small diameter of the tubing and the optimized flow rate, together with precise electronic control, enable sufficient separation of samples and wash. By allowing col-orimetric reactions to go only partially to completion, high throughput rates are possible, up to 300 h^{-1}. Although reagent consumption is low compared with the older segmented flow methods, the newer systems are even more economical than FIA. Two areas particularly suited to FIA are stopped-flow analysis as used in some immunoassays, and enzymatic analyses.

FIA systems are manufactured by:

- Burkard Scientific: http://www.burkardscientific.co.uk/Analytical/Systems_Analysers_FIAflo2000.htm
- ChemLab Instruments Ltd: http://home-1.worldonline.nl/~chemlab/

(The ChemLab instrument is in use at the Department of Soil Science, the University of Reading: http://www.rdg.ac.uk/soil/SoilSci/FACILITIES/analytical.html)

- Foss Tecator: the FIAstar® 5012 System: http://www.foss.dk/foss.asp
- Note: a useful list of scientific equipment suppliers is available at: http://chem.yonsei.ac.kr/~lsk/company.html

We will only be dealing with segmented-flow methodology in this manual, but there are sure to be equivalent FIA methods available.

Batch Size

The total number of samples will be determined by the experimental design (above), but the batch size should be chosen to suit the equipment used to process the samples ready for analysis. There will be a maximum load for the boiling units, heating blocks and shakers etc., so forward planning will opti-mize throughput. If an herbage batch size for a researcher is 80, it may be advantageous if he added half of the following batch in order to bring the batch for analysis to 120. This is because of the larger capacity of the heating block and the four available autoanalyser sample trays. Another approach would be to accumulate, say 240, samples and digest in two batches. The first batch of digested sample solutions could then be analysed while the second batch is digesting, or they could all be stored until a suitable time for autoanalysis. The main point to make is that the analytical laboratory needs to inform users well in advance of the best protocol for submitting samples. This includes other factors such as amount of sample required, recommended drying procedure, labelling, when to bring them in, and what authorized cost code is to be used in charging for the work. The question of prioritizing samples for certain users and situations in which queue jumping is allowed should also be addressed.

Sampling Protocol

In this section we will consider some precautions necessary for the sampling of various materials before analysis. In general, samples should be representative of the bulk samples from which they were taken.

> It is shown that the variation associated with field sampling is 5 to 10 times greater than that associated with laboratory procedure. It would therefore be better to increase the number of core samples taken from the field than try to improve the accuracy of the analytical methods if the precision of the results from our field experiments is to be improved.
>
> (Allen and Whitfield, 1964).

Enough core samples should be taken throughout the field or mass of material to give a representative bulk sample. This may weigh several kilograms, so should be thoroughly mixed and sub-sampled, perhaps on site, to obtain a truly homogeneous sample of a size suitable for processing.

Galvanized sampling tools should not be used for trace element analysis. Usually from 20 to 25 cores are taken in a 'W' pattern across the whole area. An alternative approach is to traverse the whole area in a zig-zag manner, sampling at random along different sections of the area (Scott *et al.*, 1971). The cores should be broken up and mixed well in a bucket, then about 200 g retained in a labelled polythene bag.

Soils

With soil sampling from agricultural fields, it is usual to avoid any small patches of different soil (e.g. boggy or very stony); dung/urine patches, gateways and headlands should also be excluded. Large areas within the field that have had a different manuring/fertilizing history should be sampled separately. An auger, bulb-planter or trowel should be used to remove a core from an appropriate depth of 7.5 cm for grassland and 15 cm for arable.

Stones and plant debris should be discarded. Sampling should be avoided after heavy rain or in time of drought. Sampling should also be avoided for

P, K or Mg analysis for 8 weeks after applying fertilizer, 12 weeks after manure or slurry, or 12 months for pH determination after liming. Further details are available from the Potash Development Association (PDA, 1999c). If the soil is to be analysed for nitrate, it should be kept moist in a grip-top polythene bag and placed in ice as soon as possible before transport to the laboratory. Unless analysed immediately, which is unlikely, it should be frozen until a convenient time for analysis. This is to arrest microbial metabolism causing denitrification (conversion of nitrate by reduction to ammonium nitrogen and nitrous oxide gases). Biological activity and other problems have been discussed by Cresser (1990).

Composts

Composts can be made from most biodegradable materials, and could derive from many unusual sources. If it originates from municipal solid waste, however, care should be taken that no toxic and non-degradable materials remain after the supplier's separation processes. Small pieces of brick and concrete, glass and plastic (inerts), lead residues from old car batteries and cadmium from electroplated items are possible. A useful work on specifications and recommended chemical analyses of composts is the book by Bertoldi *et al.*, 1987.

The analyses specifying the compost include:

ammoniacal nitrogen	nitrate-nitrogen
calcium	nitrogen
carbon	organic matter (ignition 450–600°C)
C:N ratio	particle size
conductivity	pH
heavy metals	phosphorus
inerts	potassium
magnesium	total solids
moisture	

Feeds

Bagged feeds

Instructions can be found in the AOAC Official Methods of Analysis (Padmore, 1990, p. 69). A pointed corer consisting of a single or double tube, or slotted tube and rod, is used to remove a diagonal core from end to end of the horizontal bag. Bulk feeds should have ten or more cores from different regions. The sample should be stored in such a way that deterioration and change in composition are prevented (BS 5766, 1979).

Silage, hay and haylage

A suitable corer is needed to remove sample cores from within clamps and stacks. A motor driven corer is used in some research institutes, but is rare in other establishments. One of the first designed for research work was that of Alexander (1960). His design was just 183 cm in length, and not long enough for the depth of the average farm clamp today. We designed a three-section clamp in stainless steel to resist corrosion by the volatile fatty acids in silage (Faithfull, 1997). This is clipped inside a wooden box and will fit into the boot of a car. Table 1.1 compares the two corers. Construction details are shown in Fig. 1.2.

Table 1.1. Comparison of Alexander pattern and modified design of silage corer.

Property	Alexander pattern	Modified pattern	Reason
Assembled length (cm)	182.9	281.5	Greater depth required
Number of sections	2	3	Ease of handling
Material	Mild steel	Stainless steel	Avoid corrosion products contaminating sample

To sample the clamp, make two cuts in the membrane about 3 cm long in the form of a cross. Insert the tommy-bar into the corer, thrust the corer down vertically, and finish with a twisting action. Pull up and thrust down and twist again, repeating until the corer is full. Great care should be taken not to hit the concrete base of the clamp, as this will buckle the cutting edge. A penetration of about 38 cm was needed to fill the 15.7 cm long corer tube because of the greater compaction in the tube. The sample is removed from the tube using the tommy-bar and immediately placed in a labelled grip-top polythene bag. The middle and top bar sections are added to reach greater depths. They are secured with cross-pins held in place with insulation tape.

Sampling positions

Alexander (1960) commented on the distortion of the horizontal layers in the physical structure of the silage clamp (Fig. 1.3), and concluded that the most likely points to be representative of the whole pit would be the mid-points of the half-diagonals (Fig. 1.4). A vertical core through the centre of the clamp would include more of the top layer, which would have wilted longer, than the lower layers. Conversely, a core taken near the edge of the clamp would include relatively more of the lowest, moister layer. A core through the half diagonals would be more representative of each layer, although the optimum position might need to be determined by a more careful examination of the geometry of the clamp.

Grass and herbage species

It is vital that sufficient weight of sample is taken for the planned analyses, extra being added in case further unforeseen tests are required. Plant materials

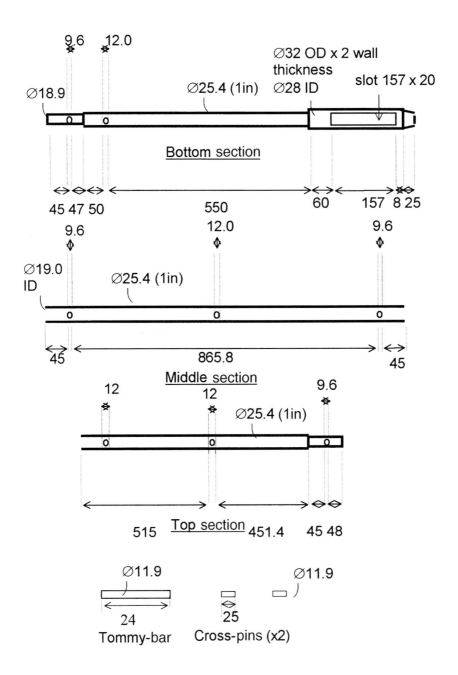

Fig. 1.2. (a) Stainless still silage corer. Units in mm (and inches (in) when appropriate).

(b)

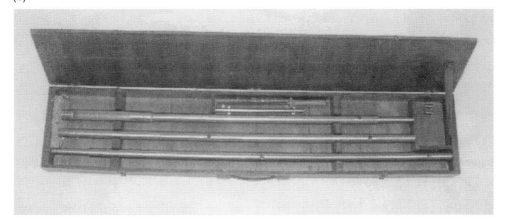

(c)

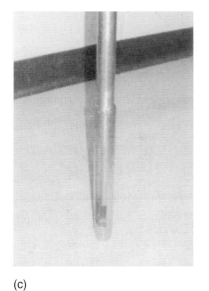

(d)

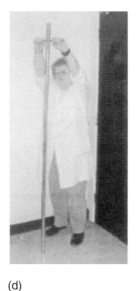

(e)

Fig. 1.2. (b) Stainless steel silage corer in wooden carrying case. From the top: file to sharpen cutting edge; tommy bar; bottom section with corer; middle section; top section. (c) Stainless steel silage corer; close-up of bottom end with corer. (d) Stainless steel silage corer; bottom and middle sections. (e) Fully assembled silage corer with metre rule.

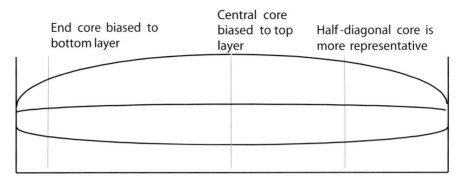

Fig. 1.3. Effect of layer structure on sample core bias.

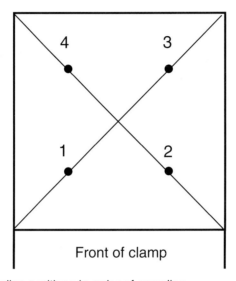

Fig. 1.4. Sampling positions in order of sampling.

are high in moisture content, and young growth could lose 85% of its fresh weight after drying (Wilman and Wright, 1978). Contamination by soil should be carefully avoided. In animal nutrition studies, however, ingestion of some soil adhering to forage leaves and stems should be considered as normal for herbivores, and thus be taken into account when assessing the mineral and trace-element status of the forage. It is normal to allow 2 weeks between grazing and sampling to avoid contamination by trampling. Washing foliage should be kept to a minimum to reduce leaching, and large smooth leaves can be wiped with a damp cloth. Atmospheric deposition immediately before sampling should be considered, especially if within 10 miles downwind of a coastal region. An assessment of the degree of contamination can be obtained from the level of titanium in dry matter. If this exceeds 10 μg g^{-1}, it can be considered as contaminated (Berrow, 1988).

Some plant species possess a high moisture content, little structural fibre, and are very delicate. Such a species is chickweed (*Stellaria media* (L.) Vill.)

with about 91.3% moisture (Derrick *et al.*, 1993). When this is thawed after being stored in a freezer, most of this moisture exudes out and so various soluble components will be lost unless poured back over the foliage before drying. Even before thawing, it forms ice crystals within the polythene sample bag, so these should be added to the sample if freeze-drying.

Plant components

The chemical composition varies between roots, stem and leaf. For a whole-plant analysis, it is essential that no root fibres are left in the ground, and that no other parts snap off and are left out of the sample. As much material as possible should be collected to minimize errors from variations in hetero-geneity. If sampling at the pollen shedding stage, the heads should be con-tained in paper or polythene bags to collect pollen and anthers (Wilman and Altimimi, 1982). It is possible to separate many types of plant components. The variation of chemical nutrients within these components and the change in them with maturity is relevant to animal nutrition. It could also influence the cutting height of crops for conservation. Some ryegrass components are shown in Fig. 1.5.

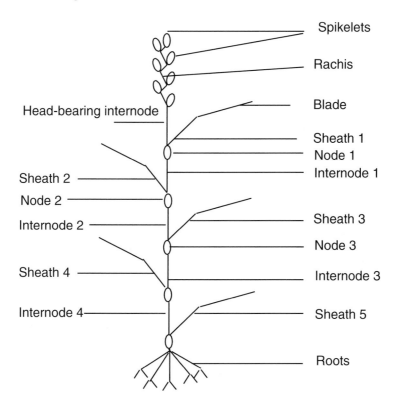

Fig. 1.5. Components of a typical ryegrass plant (*Lolium perenne* L.), adapted from Wilman and Altimimi (1982).

Microbiological analysis

Samples taken for chemical analysis may also be used for microbiological analysis. This may be the case for silage samples, when harmful clostridia could spoil the beneficial fermentation of *Lactobacillus*. It is therefore essential that the treatment of the sample immediately after collection should both prevent the further growth of the microbial species present and protect from the ingress of any harmful microorganisms or fungal spores. Although biased towards food samples, *Microbiology for the Analytical Chemist* by R.K. Dart (1996) is a helpful publication.

Biological substances

Such samples include milk, blood, urine and faeces. Most samples will only need to be placed in an ice-box after sampling, this will help to prevent degradation and oxidation of sensitive compounds like vitamin E (tocopherol). The treatment may depend on the analyte to be measured, so it is essential to study the published sampling protocol before arriving to take the sample. Blood may need to be collected in a heparin tube if plasma is to be later prepared by centrifugation. The blood should be mixed with the heparin by slowly inverting several times, but never vigorously shaken. A heparin tube is not required before centrifugation for serum preparation. Samples may be kept for several months in a freezer at $-20°C$, but for longer than 6 months at $-80°C$. If semen is to retain its activity, it should be kept in liquid nitrogen.

Fertilizers

For the sampling of fertilizers, consult Johnson (1990b), also refer to Chapter 2 'Sub-sampling' and Chapter 6.

2   Sample Preparation

Pre-treatment of Samples and Contamination

Care must be taken to avoid contamination of samples before analysis. Common causes of contamination are:

- lime or fertilizer blowing on to plots from adjacent plots/fields,
- use of tap water instead of deionized or distilled water when washing plants or extracting soluble components,
- failure to wash earth from roots thoroughly before analysis.

Trace Element Analysis

Extreme care is necessary in trace element analysis. Before use, polythene containers for storing sample and standard solutions should be washed successively with:

- 0.05 M EDTA (14.63 g EDTA + 4.0 g NaOH l^{-1})
- H$_2$O, deionized
- 1.5 M HNO$_3$
- H$_2$O, triply deionized or distilled (Adriano *et al.*, 1971).

Earth dust must be rigorously excluded and gently washed from foliage if necessary.

© 2002 CAB *International. Methods in Agricultural Chemical Analysis: a Practical Handbook* (N.T. Faithfull)

Solutions and leachates for analysis must be particle free, and should therefore be centrifuged in polypropylene tubes and *not* filtered unless this is specified in the methodology.

Sub-sampling

A bulk sample should be thoroughly mixed until homogeneous, then a sub-sample taken. There are two main ways to achieve this when dealing with solid samples. First, there is the manual cone and quartering method. A spoon-shaped spatula is used to take portions randomly from the bulk sample, which are then transferred to a clean surface to form a new conical pile. Each successive portion is poured on to the apex of the cone until the entire heap has been transferred. The cone is then flattened, divided into quarters, and opposite quarters removed. These are mixed to form a smaller conical pile, and again quartered. This is repeated until a sample of suitable weight is obtained (Jeffery *et al.*, 1989, p. 154; MAFF/ADAS, 1986, p. 2). A variation on this method is to place the sample in the centre of a square sheet of paper and thoroughly mix by alternately lifting opposite corners of the paper so as to roll the sample particles towards the centre, rather than allowing them to slide. The pile is made approximately circular and quartered as above (Triebold, 1946).

Second, easily flowing granules or powder may be riffled. This is recommended for fertilizers (Johnson, 1990b). Riffle boxes (sample dividers or splitters) are available to BS812 and BS1377 from:

A.J. Cope & Son Ltd, 11/12 The Oval, Hackney Road, London E2 9DU, UK

Tel: +44 (0) 20 7729 2405 Fax: +44 (0) 20 7729 2657

E-mail: marketing@ajcope.co.uk

also from Merck [VWR International at http://www.merckeurolab.ltd.uk/], and larger ones from Fritsch. A rotary cone type sample divider would probably be too sophisticated for fertilizers, plants and soils, with the simpler manually operated types being adequate. Cresser (1990) recommends a chute splitter or spinning riffler for environmental samples. Pascall Engineering Co. Ltd market the Rotary Wholestream,

(see: http://www.pascalleng.co.uk/sampling/representative_ sampling.htm).

This divider is intended mainly for providing samples for chemical analysis. The samples are taken from a moving stream of powder by a set of rotating stainless steel containers, the powder being fed from an adjustable hopper onto a vibrating feeder. They also make the Centrifugal, which is used mainly for seed samples and is used by the Official Seed Testing Station of England and Wales and by seed merchants and seed associations throughout the world. The Rules for Seed Testing, issued by the International Seed Testing Association (http://www.seedtest.org), give details of the unit and its use. Gross samples can be divided in seconds, and the model is suitable for all but the chaffiest of seeds.

Drying Techniques

It is important to find out the correct drying method for the nature of sample and the type of analysis to be carried out. If samples are presented for the analysis of water-soluble carbohydrates having been dried for 24 h at 100 °C, it will be a waste of time as the sugars will be partially degraded. This also applies to cell-wall analysis by the neutral detergent fibre procedure. The original fresh herbage sample will have wilted, unless frozen, so the outcome could be disastrous. Often a compromise will be necessary if both above ambient temperature and time degrade the material. Thus the choice of drying technique will be between a low temperature for a longer period or a higher temperature for a shorter period. The decision may have to be in line with the conditions published in current journals for similar experiments, and these references may be cited to justify the choice.

Air-drying

Air-drying is the usual method for soils. Large numbers of samples could be placed in cardboard or expanded polystyrene trays on metal shelving units in a ventilated warm room. A space-heater could be used to raise the temperature to no more than 30°C.

Oven-drying

Fresh plant material is generally dried in a forced-draught oven. Samples may be placed in aluminium trays with mesh in the base to allow circulation of air. Fairly dry herbage can be placed first into labelled manilla envelopes or brown paper bags, being careful not to let them touch the interior surfaces of the oven. For *in vitro* digestibility, the drying time at 80°C should not exceed 6 h. To avoid losses when determining fluoride and selenium, the temperature should not exceed 50°C, and for boron 60°C (in an unlined tray). Although a short drying time of 2 h at 102°C has been given for water-soluble carbohydrates (MAFF/ADAS, 1986b), we would only recommend freeze-drying (see below). Suggested drying conditions are given in Table 2.1. It should be noted that sample drying conditions are sometimes different from those used for dry-matter determinations, which are often more severe and for which a sub-sample is taken.

Vacuum oven

This is one of the recommended drying methods for moisture in animal feed (Padmore, 1990, p. 69). About 2 g animal feed is dried to constant weight at 95–100°C under a pressure of $\leqslant 100$ mmHg for about 5 h. A high molasses content requires $\leqslant 50$ mmHg at $\leqslant 70$°C. Vacuum ovens are ideal for drying

Table 2.1. Some drying times for various feeds.

Reference	Sample	Drying time (h)	Temperature (°C)
MAFF/ADAS, 1986, p. 4	Herbage, hay (silage dry matter)	18	100±2
MAFF/ADAS, 1986, p. 4	Brassicas	18	100±2
MAFF/ADAS, 1986, p. 4	Root crops (carrots, swedes, etc.)	48	60
MAFF/ADAS, 1986, p. 4	Potatoes, artichokes	24	60
		plus 18	100±2
MAFF/ADAS, 1986, p. 4	Cereal grains	40	100±2
Wallinga *et al.*, 1995	Herbage	24	70
Isaac, 1990, p. 41	Herbage	24	80
Byrne, 1979	Herbage	16	95

products that are heat labile at low temperatures. For dietary fibre analysis, drying at 100–105°C may cause Maillard reaction products which analyse as lignin. It is recommended that a vacuum oven at 60°C or freeze-drying is used (Southgate, 1995, p. 46).

Makers of vacuum ovens include Gallenkamp (from Fisher), Heraeus GmbH & Co. KG, Jouan, and Townson and Mercer (see http://www. sanyogallenkamp.com; http://www.heraeus.com; http://www.jouaninc.com; http:// www.townson-mercer.co.uk).

Freeze-drying

This is one of the best methods for drying sensitive materials, but has relatively little mention in the literature. It is the only way water can be almost completely removed from tissue or organic material with minimal damage to the cell structure. The fresh herbage is first deep frozen as soon after harvesting as possible. It is transferred to the freeze-dryer chamber, and the methodology used is described below. A vacuum is applied, and a controlled supply of heat may be provided. This is to allow the ice to sublimate or evaporate, but never to melt. The extracted water vapour condenses on the surface of the refrigerated chamber at about –40°C. A small amount of water vapour escapes condensation and passes out to the vacuum pump and through the oil reservoir, thence through an oil mist filter to the atmosphere. If the oil is not hot, the water vapour will condense in the oil and sink to the bottom. The vacuum is measured using a Pirani gauge, with readout on a meter; this may have a separate electrical switch.

Notes:

A freeze-dryer must be purchased that either has a large built-in chamber, or to which a chamber can be attached. Some smaller ones are mainly for small-scale work when multiple samples are held in glass flasks or ampoules, which are attached to a manifold equipped with isolation valves.

It is vital to maintain clean contacts on the connection fitting of the Pirani gauge, because it is very sensitive to resistivity changes due to tarnishing or dust.

Freeze-drying methodology

The freeze-dryer is switched on and the pump started in order to warm up the oil within the pump housing. This is to prevent condensation of moisture in the oil. The gas ballast valve is also opened for the first half of the drying process, when most of the moisture is removed, to purge out any moisture from the oil. One manufacturer (ChemLab) recommends leaving the ballast valve open for the whole drying process. Although this will keep the accumulation of water in the oil to a minimum, it will mean the ultimate vacuum possible with the pump will not be reached, and samples will have slightly more residual moisture. When the condenser temperature indicator reads less than −30°C, the frozen sample may be loaded into the chamber. Samples may be placed in trays, paper bags or microporous bread bags, but preferably not in polythene bags, which could hinder the evaporation process. Samples of a lumpy consistency should be broken up while still frozen to speed the evaporation process. The chamber may have rubber seals which need greasing. The minimum amount of silicone grease should be applied, and the seals should be wiped scrupulously clean before the application, as any particles of sample material adhering to the seals will allow ingress of air owing to poor sealing at that point. With the chamber lid or door in place, the drain valve should be closed, and the vacuum valve opened. When the Pirani gauge reads 66.5 Pa (500 millitorr or 0.5 mmHg), the heater may be switched on. The drying time will depend on the nature and water content of the samples, but 2–4 days is normal. The condensation chamber will have a certain capacity, perhaps 3 l, so the total sample water content should not exceed this, and there is a loss in efficiency if about two-thirds of this value is exceeded.

At the end of the process, when the pressure has been about 13.3 Pa (100 millitorr or 0.1 mmHg) for several hours, the isolation valve is closed, the drain tap opened to allow ingress of air, and the defrost switch turned on. The lid may then be removed and the samples checked for dryness. Any larger samples should be inspected, to ensure there are no remaining areas of ice at the centre. The freeze-dried samples should be stored in a desiccator before milling, which should be done as soon as possible. Freeze-dried samples are somewhat more sticky than oven-dried ones, and the crisper they are, the better they mill. After milling, the samples should be stored in airtight sample tubes or grip-top polythene bags to prevent rehydration and fungal attack. Samples will not be as dry as oven-dry material, and will typically contain between 3% and 10% moisture depending on the sample. When results from the analysis of freeze-dried material have to be expressed in terms of oven-dry matter, a sub-sample must be taken at the time of weighing for a separate oven-dry matter determination. This will enable a correction for residual moisture to be made.

Microporous bags can be obtained from Cryovac at the website of Sealed Air Corporation:

http://www.sealedair.com/products/food/bakery_fs.htm
Websites of freeze-dryer manufacturers/suppliers are:
Virtis: http://www.virtis.com/
ChemLab: http://home-1.worldonline.nl/~chemlab/
CHRIST: http://www.phscientific.co.uk/html/
Heto: http://www.heto-holten.com/camel.htm

Desiccation

Although infrequently used, an alternative method for drying samples at room temperature is in a vacuum desiccator. These used to be made of glass with a possible risk of implosion. Modern ones are made from polycarbonate or polypropylene base with a polycarbonate cover, and are cheaper than the glass equivalent. The main limitation is that they are for room temperature use only, and not for use with organic solvents or vapours. An efficient water pump should be adequate, however, a guard-tube containing desiccant should be inserted between the pump and the desiccator. Manufacturers include Kartell and Nalgene.
Nalgene: http://nalgenelab.nalgenunc.com/

Milling, Grinding and Homogenization

Animal tissue is often blended in a high-speed blender until completely homogeneous. For trace element work, solid samples should not be ground in a mill constructed with materials containing the elements to be determined, such as iron, chromium and manganese. In this case, a mortar grinder (mortar and pestle mill) or ball mill would be suitable. The former may be constructed of agate, with the pestle and mortar being independently motor driven (Pascall Model 00, Agate), or may consist of a vibrating ball and mortar (Fritsch Pulverisette). The ball mill may be porcelain with fused magnesium silicate balls (Pascall Model No. 9). There are centrifugal, planetary and roller type ball mills. The physical characteristics of the sample material may determine which type is best for the purpose, and the manufacturer's advice should be sought. Some manufacturers are listed below:
Glen Creston Ltd: http://www.glencreston.co.uk/
Christy: http://www.christy-norris.co.uk/
Fritsch GmbH: http://www.fritsch_lab.de/englisch/english.htm/
IKA®: http://www.ika.net/
Merck: http://www.merckeurolab.ltd.uk/
Pascall Engineering Co. Ltd: http://www.pascalleng.co.uk/Mixing.htm
Retsch GmbH & Co. KG:
http://www.retsch.de/english/zerkleinern_e.html
The fineness of grind is important and can influence the result, especially when the sample is being subjected to partial dissolution in detergent or enzyme containing solutions. A mesh of the appropriate size can usually be

inserted in the mill. Herbage is usually graded to 1 mm particle size. For available carbohydrates in cereal mixes, the sample should be ground to 0.5 mm.

In general, for small dry samples, a micro hammer-cutter mill will tackle anything from cotton to small rocks. For more than 100 g dry herbage, a larger cutter or knife mill will be more efficient. We use a Christy-Norris 20.3 cm (8 in) cutter mill with a 1.47 kW (2 hp) 415 V (three-phase) motor. A 12.7 cm cross beater hammer mill at approximately 13,000 rpm is also suitable. If the receiving container is a cotton bag, it should be turned inside out and shaken between samples. An 18 × 36 cm grill of 2.5 cm wire mesh set into the bench in front of the mill and connected to a suction fan (415 V, 2.5 A, 1400 rpm, 1.1 kW) via ducting through an outside wall, removes the dust at source. It is essential to clean the mill between samples to prevent cross-contamination, and a paint brush and vacuum nozzle are used. However, if milling samples weighing about 500 g, and the component to be measured only differs by a maximum of 0.5% between samples, then a residue of 5 g in the milling chamber will only affect results by 0.005%. If results are given to 0.1%, the tedious cleaning process might be considered unnecessary.

The lignified and cutinized tissues of cereal grains need the more vigorous disintegration of ball-milling to produce a homogeneous sample. Prolonged ball-milling, however, can depolymerize cellulose, therefore wet ball-milling in an organic solvent or suitable extractant is recommended (Southgate, 1995, p. 47).

Freezer mill

For extra sensitive or rubbery samples, a freezer mill is available. This uses liquid nitrogen at −195.8°C which renders most ductile or elastic substances friable, and is suitable for those with a low melting point or which are unstable at room temperature. The sample is placed in a polycarbonate tubular sample container with a stainless steel impactor and closed with two end caps. It is immersed in liquid nitrogen and a magnetic field oscillates the impactor against the end caps to powder the sample. Samples of up to 3 ml can be ground in less than 4 min, while others wait in a separate compartment in the milling bath. A full day's operation may require 20 l of liquid nitrogen.

Homogenization

This may find application in several areas. The first example is the homogenization of animal tissues in a high-speed blender, which enables a homogeneous sample to be obtained for subsequent analysis. This is used, for example, in the analysis of arsenic or copper in liver (Ross, 1990). A second area is the extraction of volatile fatty acids from silage. Typically, 10 g fresh silage is homogenized for between 1 and 10 min with 100 ml water in a blender before filtration (Lessard *et al.*, 1961). The last area is the dry homogenization of linseed seeds prior to oil determination. These resist crushing and so about 4 g of seeds are homogenized at 11,500–13,000 rpm for 2 min.

There are several types of blender. One of the most popular is the Waring blender, operating on a similar principle to domestic jug-type food blenders. Containers are available in glass, stainless steel and polycarbonate. It should be noted that the working capacities of the containers are a maximum of 70%, and a minimum of 10% nominal capacity. Accessory containers enable volumes as low as 12 ml to be handled. Silage is homogenized in a Waring type blender.

There are also the dispersing shaft type homogenizers, which may be hand held or stand mounted. The shaft has a tip with teeth rotating within a fixed crenated stator, which imparts impact, shock, shearing and cavitation effects. Working volumes as low as 0.03 ml (PRO20 Homogenizer) can be handled. The materials, however, should be free-flowing, and usually suspended in a liquid. For solid materials, like seeds, a blade rather than a dispersing tool is required. PRO™ market a Safety-Seal® Chamber Assembly with a 25.4-mm blade which can handle a minimum of 10 ml (supplied by Radleys). Status homogenizers can be equipped with their AX60 Analytical Mill attachment (supplied by Philip Harris Scientific). This has a cooling jacket that can be used with liquid nitrogen for temperature sensitive samples.

Manufacturers' websites:
Büchi Labortechnik AG: http://www.buchi.com/
Fisherbrand: http://www.fisher.co.uk/
IKA®: http://www.ika.net/
http://www.labworld-online.com/ika/index1.html
Kinematica (Polytron®): Kinematica AG at http://www.kinematica.ch/
Tel.: +41 41 2501257 Fax: +41 41 2501460
Supplied by Philip Harris (RossLab plc)
http://www.phscientific.co.uk/html/
PRO Scientific Inc: http://www.proscientific.com/
Radleys (R.B. Radley & Co. Ltd): http://www.radleys.co.uk/

Storage of milled samples

Once milled, the samples should be stored in air-tight containers and kept in a cool place away from direct sunlight. Powders are suitably stored in 50 × 25 mm glass specimen tubes capped with polythene push-in closures. They may be handled in aluminium or polycarbonate freezer trays. Two sizes of trays are useful – a larger size holding 120 tubes (10 × 12) is suitable for oven redrying before weighing. A smaller tray holding 25 (5 × 5) will fit most desiccators for temporary storage after redrying for subsequent weighing.

The sample tubes should be numbered consecutively from 1 up. Any plot codes, identifying letters, etc., should be kept by the person submitting the samples for later interpretation. This simplifies the sample labelling and record keeping of the analytical laboratory. (*Note*: Unique batch/sample numbers are required for UKAS accreditation.)

When sub-sampling from a kilogram or more of milled herbage samples or sieved soils, it would be wise to store the remainder of those giving low, medium and high values for future use as reference samples. These can be included with each batch of similar samples, and thus any excessive standard deviation from the mean (obtained by repeated analyses over a period of time) will indicate that an error has arisen in the analytical procedure. A protocol should therefore be established that if one (or more) of the low, medium and high control samples included with the sample batch gives a result lying outside of, say, ± 2s (where s = standard deviation), the whole analytical procedure should be repeated after checking from where the error could have arisen. This is discussed in greater depth in Chapter 12.

3 Weighing and Dispensing

Weighing Errors

There are various sources of error that can occur when weighing samples for analysis.

Correction of weighings to 'in vacuo'

If, as is usual, the sample has a lower density than the stainless steel balance calibration weight, the buoyancy effect of air on the sample mass means that a litre of water would indicate a weight of 1.05 g less than expected. Where weights of sample components are expressed as percentages or ratios, this error almost reduces to zero. This correction is usually ignored and considered well within acceptable experimental error (Jeffery *et al.*, 1989, p. 76).

Incorrect calibration of the balance

Between periods of servicing of the balance, it is wise to check the accuracy with a calibration weight. Some balances incorporate a self-calibration facility.

Static charge

If a glass vessel is cleaned by wiping or brushing, a static charge may build up. For a vessel of 150 cm², the force on the balance pan could amount to 100 mg. Clearly, when weighing samples of less than a gram, this is an unacceptable error of over 10%. Static may also be present on the person weighing or on the sample particles; milled herbage can jump from the spatula blade on to the walls of the weighing vessel. The apparent weight often alters as the spatula is lowered into the weighing container. The final reading should therefore only be taken after withdrawal of the spatula and when the draught shielding door has been closed. Electronic anti-static devices are available, but as they usually incorporate a fan, it is necessary to position them carefully to avoid the effect of the draught.

Convection currents

It is essential that both samples and crucibles have cooled to room temperature in a desiccator before weighing. This may take 30–40 min.

Absorption of moisture by the sample

Although dried herbage samples are kept in a desiccator before weighing, it is possible for samples to absorb moisture from the atmosphere during weighing (Faithfull, 1970). This arises from the repeated removal and replacement of the desiccator lid. This was investigated using three types of sample: grass, barley and faeces. Using a large desiccator holding 80 samples, the lid would be removed that number of times over a 2-h period. The absorption of moisture by the last sample to be weighed amounted to 0.95% for grass, 0.83% for faeces and 0.77% for barley. This effect can be reduced to about 0.1% by using smaller desiccators holding about 12 sample tubes.

The sample will continue to absorb moisture while on the balance pan, the initial rate being about 0.01% min^{-1}. The use of a well-balanced spatula (e.g. a wooden handled 75 mm stainless steel-bladed palette knife) and glass weighing funnel will speed the weighing process and reduce moisture absorption.

Absorption of moisture by the sample container

Glass and porcelain are particularly susceptible to adsorption of atmospheric moisture on exposed surfaces. Containers to be heated in gravimetric procedures (e.g. oven-dry matter or ash content) should therefore be pre-heated to the same temperature as that procedure and cooled in a desiccator before measuring the tare weight.

Dispensing Errors

Dispensing errors can arise from the use of inappropriate or inaccurate equipment. Measuring cylinders are suitable for making up reagents, but are not accurate enough for the dilution of sample solutions. For the latter, a pipette and graduated (volumetric) flask are used. The accuracy of these may be checked by weighing the dispensed or contained amount of water. Flasks are calibrated on manufacture at 20°C. If a 1-l borosilicate glass measuring flask is used at 15°C, the contraction of the glass wall increases the volume by 0.05 ml, thus a correction of –0.05 ml is required. The water has itself contracted by 0.84 ml, so an additional correction of +0.84 ml should be added, making the total correction +0.79 ml, or +0.079%. Although this is an acceptable error, when combined with other sources of error, the maximum possible error can be surprisingly high. Thus each source of error should be minimized as far as is practicable. Volumetric glassware is available in Class A and Class B qualities. A Class A 10-ml bulb (one-mark) pipette has a tolerance of ±0.020 ml, and a Class B ±0.040 ml. Class B is adequate for routine agricultural chemical analysis.

Bulb, or transfer pipettes, are usually made to deliver a stated volume of liquid under standard conditions of temperature and with a draining time of 15 s while the tip is in contact with the wall of the receiving vessel. Previously, the tip should be touched against the wall of the container from which the liquid has been aspirated in order to allow any adhering droplet to drain away. A pipette filler should be used to avoid the danger of liquid entering the mouth when the unsafe mouth suction technique is used. Graduated pipettes with straight sides may deliver a volume from zero at the top to any graduation line, or from a graduation line to zero at the jet tip. Some are blow-out pipettes which require the last drop to be blown out from the tip, and these are indicated by a white or etched ring near the top of the pipette. Normal bulb pipettes should never be blown out to try and save time. Bulb pipettes are graduated to BS1583 and graduated pipettes to BS700 and ISO835, and most are colour coded. The latter are divided into types as given in Table 3.1.

Bottle top dispensers

Bottle top dispensers are invaluable for the repetitive measurement of a certain volume of reagent into sample containers for extraction. For example, they find extensive use in soil analysis for dispensing the extracting reagents for phosphate, potassium and magnesium. If the volume setting is adjustable, it is essential to check the amount delivered by weighing the water. Careful priming should ensure that there are no trapped air bubbles.

Some manufacturers are:

Bibby Sterilin Ltd: http://www.bibby-sterilin.co.uk/
Brand GmbH & Co. KG: http://www.brand.de/
Eppendorf AG: http://www.eppendorf.com/

Jencons (Scientific Ltd)/Zippette: http://www.jencons.co.uk/
John Poulten Ltd/Ultra Volac:
http://www.pdd.co.uk/experience/medical/volac.htm and
http://www.labpages.com/std_home/page0107.html

Table 3.1. Various types of graduated pipette.

Type 1	Calibrated for delivery (EX) from zero at the top to any graduation line down to the shoulder
Type 2	Calibrated for delivery (EX) from any graduation line down to zero at the jet
Type 3	Calibrated for delivery (EX) from zero at the top to any graduation line down to the jet
Type 4	Calibrated to deliver from zero at the top down to the jet with the last drop expelled by blowing

Syringe pipettes

Syringe or micropipettors can be of two types: positive displacement or air displacement. In the former type the liquid comes into direct contact with the piston, which may lead to carry-over from one sample to the next, albeit usually negligible. This would not matter if the same reagent solution were being dispensed. The latter type, however, is usually used. They have either fixed or adjustable ranges and are available from 1 µl to 10 ml. Micropipettors with disposable tips are useful for dispensing the extracted and filtered soil solutions. The delivered volume should be checked as above, and care taken to use the correct technique. With the pipette in an upright position, the push-button should be slowly depressed until the first resistance is felt (first stop position). With the tip well immersed in the liquid to be dispensed, the push-button is slowly released. When aspirating the solution, no air should be admitted by exposing the tip above the liquid surface. If this happens, liquid will contaminate the piston chamber, which should be cleaned before further use. The liquid is delivered by slowly depressing the push-button until the first stop. After a couple of seconds, press the plunger to blow out the droplet to empty the tip.

When checking the volume of water delivered by weighing, Table 3.2 will enable a graph to be plotted and the volume at the exact temperature of measurement determined.

Table 3.2. Volume of 1 g water at temperatures between 10°C and 30°C.

°C	Vol. (ml)	°C	Vol. (ml)
10.00	1.0013	22.00	1.0033
12.00	1.0015	24.00	1.0037
14.00	1.0017	26.00	1.0044
16.00	1.0021	28.00	1.0047
18.00	1.0023	30.00	1.0053
20.00	1.0027		

4 Acid-digestion, Ashing and Extraction Procedures

The actual analytical methods will be detailed in the appropriate chapters, but here we will just comment on the techniques involved.

Acid-digestion and Washing

Acid-digestion of soils

There are three main reasons for digesting soils in hot acid – to determine the organic carbon content, to extract mineral elements for their total content, and to determine total nitrogen by the Kjeldahl digestion.

The first is called Tinsley's wet combustion (Tinsley, 1950), and uses a highly corrosive mixture of sodium dichromate, and concentrated perchloric and sulphuric acids. For undergraduate practical classes, the safer loss on ignition method might be considered more appropriate.

The second reason for acid-digestion is the determination of the total soil elemental content of, e.g. potassium, phosphorus or trace elements. This is seldom done for potassium in normal soil samples, mainly because 'the total K in soils is of no value as an index to the availability of K to plants, nor is it always of value in tracing the movement or accumulation of applied fertilizer K' (Pratt, 1965). The unreactive soil phosphorus is obtained by subtracting the naturally leached reactive phosphorus from the total phosphorus, and a method for determining the latter by extraction with sulphuric acid and potassium persulphate is cited by Turner and Haygarth (2000). They analysed the reactive phosphate by flow injection analysis using a Tecator 5020 with autosampler, and using Method Application ASN 60–03/83 (Tecator Ltd, Sweden). The safety aspect is an important reason for avoiding, if possible, total elemental determination in soils, because the reagents often involve hydrofluoric acid (48% m/m) and perchloric acid (60% m/m). The former causes horrific burns, possibly fatal if not treated immediately, but is necessary to dissolve the potassium-bearing silica, and the latter, necessary for completely dissolving organic matter, may cause explosions if evaporated to dryness with carbonaceous materials or metals. Alkali fusion is another method for total elements in soil.

Acid-digestion is often used with composts derived from municipal wastes, sewage and slurry, where toxic amounts of heavy metals may cause problems on the land to which they are applied. It is probably more convenient to determine total elements in soils by a benchtop X-ray fluorescence spectroscopy (XRF) instrument. This only requires the soil to be ground, and several reference standards of a similar soil. *A Reference Materials Catalogue*, Issue 5, 1999, is available from LGC's Office of Reference Materials, Queens Road, Teddington, Middlesex TW11 0LY, UK. Tel. +44 (0)20 8943 7565; Fax +44 (0)20 8943 7554.

Alkali fusion, hydrofluoric acid (HF) digestion and XRF give true *total* values as required for geochemical purposes, but digestion in *aqua regia* (see Method 5.15) gives total environmentally available concentrations, which are most meaningful for agricultural and environmental purposes. Transition metals may be more effectively extracted by using a pressured microwave digestion system such as the Anton Paar Multiwave Microwave Sample Preparation System. An example of sewage sludge analysis by this system is given at: http://www.lab123.com/app_data/mswave.htm

Total soil nitrogen

Soils mainly contain nitrogen in its reduced state such as ammonium compounds and organic amino complexes. The standard Kjeldahl technique is therefore suitable to estimate the organic (plus ammonium) nitrogen, which it does by oxidizing the organic matter in hot sulphuric acid containing a catalyst and converting the nitrogen to ammonium sulphate, which can be measured by distillation and titration, or by a colorimetric procedure. The distillation is carried out after first adding excess sodium hydroxide to the acid digest to liberate the ammonia gas from the ammonium sulphate.

$$(NH_4)_2SO_4 + 2NaOH = Na_2SO_4 + 2NH_3\uparrow + 2H_2O$$

This is distilled into a receiving flask containing boric acid indicator mixture and titrated against 0.001 M HCl. The colorimetric method using the auto-analyser is based on that used for plant materials (see below), but care should be taken that any precipitate formed does not collect in the flowcell, which must be occasionally inverted or cleared by passing a bubble of air through it. Any nitrate (and nitrite, which is usually insignificant) should be reduced to ammonium by adding salicylic acid followed by zinc dust (see Method 5.6a.i) before digestion for the autoanalysis method with a colorimetric procedure, or Devarda's alloy (see Method 5.5b.ii) before the distillation, if it is to be included to give a total nitrogen value.

Acid-digestion of plant materials

The original method for the determination of nitrogen by sulphuric acid-digestion was published by Kjeldahl in 1883 and fully described by Burns (1984). Many modifications have since been made with various catalysts and acid mixtures.

The digestions can be carried out in up to 40-place multiple heating units using specialized glassware which is commercially available; some suppliers are listed below:

Digestion systems:

Büchi Labortechnik AG: http://www.buchi.com/

Gerhardt UK Ltd: http://www.gerhardt.de/gb/kb.htm

Foss (Digestor 2000 System): http://www.foss.dk/foss.asp

Distillation systems:

Büchi Labortechnik AG: http://www.buchi.com/

Foss (Kjeltec® 2300 Analyzer Unit): http://www.foss.dk/foss.asp

Gerhardt UK Ltd http://www.gerhardt.de/gb/vap.htm

We have devised a method enabling the digestion of up to 152 samples at a time, and with the wearing of some essential personal protective equipment (PPE), it has proved successful for over 32 years (Faithfull, 1969).

Acid-digestion unit

The major expense is the hotplate, which has to have a sufficiently large working surface area and be able to sustain a temperature of 310°C. A suitable hotplate is the Gerhardt HC 63, nominal voltage 400 VAC, 4800 W, working area 650 × 300 mm, and a maximum temperature of 400°C ± 5°C. In the UK this is available from:

C. Gerhardt UK Ltd, Unit 5, Avonbury Court, County Road, Brackley, Northants. NN13 7AX.

Tel. +44 (0) 1280 706772; Fax. +44 (0) 1280 706088

Other suitable hotplates are available from S & J Juniper & Co.:

http://www.sjjuniper.com/general_purpose.shtml

On the centre of the work surface are positioned two aluminium blocks, 440 × 100 × 100 mm (w × d × h), with the bottom surface machined flat to ensure good thermal contact with the hotplate. These are each drilled with 17 mm diameter holes to a depth of 86 mm and arranged in four rows of 19 holes. Thus each block accommodates 76 digestion tubes. These tubes are 150 mm long and 16 mm diameter, heavy wall (BS 3218) borosilicate glass rimless type; they are supplied by Fisher as TES-674-150S. The exposed areas of the work surface may be covered with a heat-resistant insulating material.

The whole unit is accommodated in a fume cupboard fitted with a scrubber unit to remove the acidic fumes before emission to the atmosphere. The constructional materials of the fume cupboard should be able to withstand the heat radiated from the hotplate and heating blocks. A digestion tube containing a 350°C thermometer with the bulb embedded in a 2-cm layer of sand occupies one hole in each block. The hotplate can be connected to the power source via a time-switch, which can be set to come on approximately 1 hour before commencement of work; this saves valuable time lost waiting for it to warm up.

Acid-digestion procedure

The acid used for the Kjeldahl digest is analytical quality concentrated sulphuric acid which contains 4 g l^{-1} selenium. This is prepared by heating a 250-ml portion of acid with 4 g selenium powder (Aldrich 20,965-1, 100 mesh) in a 1-l beaker on a hotplate in a fume cupboard, carefully stirring with a glass rod until dissolved to form a green solution (protective gloves, and safety spectacles or visor must be worn at all times when handling concentrated acids). After cooling, the solution is poured via a funnel into a glass storage bottle, and the balance of 750 ml acid added. *Note*: a dust mask should be worn when weighing selenium as it is easily absorbed by the lungs and is a possible teratogen. The beakers should be removed from the hotplate and left to cool in the fume cupboard after placing a watch glass over the top of the beaker. After cooling, the solution is poured via a funnel into a bottle or reservoir fitted with a bottle-top dispenser adjusted to 5 ml. All components with which the solution comes into contact must be resistant to concentrated sulphuric acid. *Warning*: the solution is highly corrosive and even when cold rapidly dissolves cellulosic materials. Wipe up any drips immediately with a wad of tissue and soak with plenty of running water before disposal. This is to both protect personnel involved in waste disposal and to prevent spontaneous combustion. Acid on the skin should be flooded with water for 1 min and medical advice sought for any blisters or burns; contaminated clothing should be removed and washed before reuse.

Exactly 0.1000 g milled plant sample is weighed into a glass weighing funnel and transferred to the numbered digestion tubes with the aid of a small paintbrush. The digestion tubes are held in stainless steel racks and either stoppered or covered with sheets of paper until ready for digestion. The tubes should have been previously marked with two scratch lines around the outside at the levels of 5 ml and 10 ml. The acid is dispensed carefully into each tube; if it is admitted too rapidly, fine sample powder as well as acid may be ejected from the tube. A few tubes at a time are loaded into the blocks. Some types of sample are prone to frothing, and if this occurs, it is easier to remove a few tubes and allow them to cool in their racks, rather than risk some frothing right over before they can be removed.

The most tedious aspect of the procedure is, after about an hour, to run a thin (4 mm diameter) glass rod vertically around the inside of the digestion tube in a downward spiralling motion in order to reintroduce any sample particles back into the acid. *PPE must be worn for this.* The temperature must not exceed 320°C because sulphuric acid boils at 330°C, which could cause injury; the two thermometers should be checked before the cleaning operation. (*Note*: the normal Kjeldahl procedure uses a salt such as sodium sulphate to raise the boiling point of the acid.) The samples are allowed to digest for a total of 4.25 h, when they are removed with stainless steel tongs and allowed to cool in their racks. The acid level is then adjusted dropwise with concentrated sulphuric acid to the 5-ml mark to replace any lost as fumes. Deionized water is then slowly added from a wash bottle, directing the jet down the side of the tube, up to the 10-ml mark so as to form two layers. *Note*: normally the safe way is to add concentrated sulphuric acid to water, especially when contained in a beaker – this is to prevent violent boiling. This does not happen here because of the restricted surface area and the formation of separate layers. The two layers are mixed by slowly oscillating a thin glass rod with one end flattened to form an 8–10 mm disc. Mixing should start from the junction of the layers, slowly working towards the top and bottom. The solution will contract after cooling, so the level must be again adjusted to the 10-ml mark and mixed with the rod. This final adjustment to 10 ml is best done immediately before analysis otherwise the tubes will need to be stoppered to avoid absorption of atmospheric moisture.

The advantages of this digestion technique are the large number of samples that can be processed at one time, the simple and cheap glassware involved, and the fact that the digest may be used for the subsequent determination of not only nitrogen, but calcium, magnesium, potassium, sodium, phosphorus and iron.

Summary of the indophenol blue colorimetric determination of nitrogen

To determine the nitrogen content of herbage and soils by autoanalysis, one must first carry out a Kjeldahl digest in concentrated sulphuric acid with selenium (0.4% w/v) catalyst; this converts protein nitrogen to ammonium nitrogen, as shown in Fig. 4.1. The density of the blue colour is proportional to the nitrogen content. It is measured using a spectrophotometer at a wavelength of 640 nm and the height of the peaks on a chart-recorder compared with those of known standards to obtain the nitrogen content of the original material. Protein content = %N × 6.25.

Microwave acid-digestion

The digestion of a wide range of matrices, from fish to rocks, is possible in a stainless steel pressure vessel fitted with a PTFE container. It is particularly useful for demanding trace element analyses. It was first developed by Professor Tölg, the method being described by Kotz *et al.* (1972). Pressure vessels are expensive, but digestion times can be as little as 60 s to dissolve fish tissue in nitric acid. An article comparing closed vessel microwave digestion versus conventional digestion procedures for the determination of

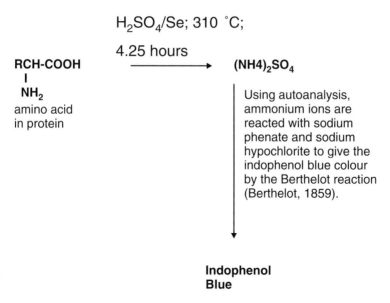

$$\text{RCH-COOH} \xrightarrow[\text{4.25 hours}]{\text{H}_2\text{SO}_4/\text{Se; 310 }^\circ\text{C;}} (\text{NH4})_2\text{SO}_4$$

RCH-COOH
|
NH₂
amino acid
in protein

Using autoanalysis, ammonium ions are reacted with sodium phenate and sodium hypochlorite to give the indophenol blue colour by the Berthelot reaction (Berthelot, 1859).

Indophenol Blue

Fig. 4.1. Colorimetric reaction converting protein nitrogen to the indophenol blue colour.

mercury in fish tissue by cold vapour AAS using a basic laboratory microwave is given by D.C. Stockton and B. Schuppener at:

http:/www.epa.gov/earth1r6/6lab/mercury.htm

Typical vessels are described at the following website:

http://www.berghofusa.com/berghof.htm

Digestion systems are supplied by CEM Corporation in the USA:

http://www.cem.com/applctns/AcdDgst.html

The UK supplier is CEM (Microwave Technology) Ltd, Unit 2 Middle Slade, Buckingham Industrial Park, Buckingham MK18 1WA, UK

Tel. +44 (0) 1280 822873; Fax. +44 (0) 1280 822342.

Their HP-500 Plus vessel system can handle 14 soil or plant digestions at a time.

Microwave systems are also used for accelerated Soxhlet extractions with reduced solvent consumption, and microwave muffle furnaces with air-exhaust for rapid ashing.

Dry ashing

Dry ashing is normally carried out in a muffle furnace. Large numbers of silica basins or crucibles take up a considerable amount of floor area within the furnace, therefore the larger the capacity the better. It may be advantageous to have two furnaces. Typical specifications would be:

interior dimensions (depth × width × height)	457 × 305 × 203 mm
volume	27 l
maximum power rating	7 kW

Suitable furnaces are available from:

> Stuart Scientific (Bibby): http://www.bibby-sterilin.com/cat/stuart/
> furnaces.htm#muffle
> Carbolite Furnaces Ltd: Aston Lane, Hope, Hope Valley, Sheffield
> S30 2RR, UK. Tel. +44 (0) 1433 620011 Fax. +44 (0) 1433
> 621198
> See also the following websites:
> http://www.catalogue.fisher.co.uk
> http://www.keison.co.uk/carbolite/carb39.htm

Ashing in a furnace is a compromise between total oxidation of carbon and some vaporization of the element of interest. When this is for trace metals, such losses only become significant with the more volatile metals such as cadmium and lead. Even some iron may be lost if chlorides are present, as ferric chloride is appreciably volatile at 450°C. A useful review of ashing biological material for the determination of trace metals was provided by Middleton and Stuckey (1953, 1954). They recommended an ashing temperature of 500–550°C (dull red heat) as the lowest temperature at which combustion can be completed in a reasonable time when the trace metal is volatile. Sometimes the sample is first moistened with sulphuric acid when lead is being determined, to convert it to lead sulphate, which is involatile below 550°C.

When ashing for trace element determination, we prefer to err on the side of caution, and recommend ashing overnight at 450°C . Sometimes an additional treatment is required such as for manganese solubilization. Ashing converts manganese salts to manganese dioxide, which is virtually insoluble in dilute acids. The ash is therefore moistened with concentrated HCl and heated carefully on a hotplate until it has fumed dry. This converts the manganese dioxide to manganous chloride:

$$MnO_2 + 4HCl = MnCl_2 + Cl_2\uparrow + 2H_2O$$

The residue is then dissolved in 0.1 M HCl in the normal way for the subsequent determination of elements by atomic absorption spectrophotometry.

It is usually best to avoid the production of flames from the sample while ashing (deflagration) as this can result in some loss of analyte. The sample should therefore be placed in the muffle at room temperature, with the chimney vent open, allowing the combustible gases to evolve without ignition as it heats up to the final ashing temperature. The vent is then closed to prevent a downdraught blowing the light ash out of the crucibles, and ashing is continued for the stated time. Dry ashing of animal tissues is problematic, and the above authors suggest using a mixture of nitric and sulphuric acids for the purpose (Middleton and Stuckey, 1954). This would be safer than mixtures involving perchloric acid, which may be explosive.

Extraction Procedures – Plant-based Materials

Other extraction procedures are used for determining: (i) oils, fats and waxes; (ii) fibre, lignin, cellulose, nitrogen-free extract and starch; (iii) *in vitro* digestibility; (iv) nitrate and water soluble carbohydrate; (v) water content in silage.

Oils, fats and waxes

All substances in feedstuffs logically belong to one of the six components or groups of a 'proximate analysis'. This concept goes back nearly 150 years to the first state agricultural research stations in Germany, and is also known as the Weende methods (Henneberg, 1864). They give crude, but useful, measurements of the components of feedstuffs, and adaptations of the original methods are used today. The components are as follows:

water	(from dry matter determination)
protein	(from nitrogen determination)
fat/oil	(also known as ether extract)
fibre	(also called crude fibre)
ash	(mineral content)
nitrogen-free extract	(subtract sum of above from 100%; it is mainly carbohydrate/starch)

Fat includes triglycerides, sterols, lecithins (phospholipids), essential oils, fat-soluble pigments such as chlorophyll, and similar substances. The AOAC recommends that anhydrous diethyl ether kept over freshly cut sodium pieces is used for the extractant (Padmore, 1990, p. 79), but we prefer to use petroleum spirit, also called light petroleum and petroleum ether, with a boiling range of 40–60°C, as it is a less hazardous solvent.

The sample should not be oven dried before analysis as this could oxidize or degrade the oil and result in too low a value. A separate sample should be taken for a dry matter determination allowing the result to be corrected to percentage fat in dry matter. The ground sample is placed in a cellulose extraction thimble of the correct size for the Soxhlet extraction glassware. The Whatman extraction thimbles are available in two thicknesses, and it is the more robust double thickness that is preferred. The size of Soxhlet flask should match that of the heating mantle recess. The 250-ml capacity recess is most appropriate, and heating units are available with three or six recesses, with the six-recess model being more economical per recess, and more suited to handle multiple samples. The heating units are specially designed to be spark free in normal operation.

The flask is pre-dried and weighed, so a flat-bottomed flask is easier to handle. After extraction, the remaining solvent is evaporated off on a boiling water bath. When there is no longer any smell of solvent, the flask is again dried in an oven (102°C), cooled and weighed. The weight of oil remaining in the flask is found by difference.

For multiple samples (c. 5 g) of seeds such as oilseed rape, another approach is to crush them and enclose them in small packets of pre-dried Whatman No. 4 filter circles, which are then stapled, labelled with a pencil and weighed. From 10 to 16 of these packets may be extracted in one large

600 ml Soxhlet with a 2-litre flask. In this case, the packets are reweighed, and the weight loss gives the crude oil content (Hughes, 1969).

Oil in compound feeds and feeding stuffs

This method published by MAFF (1993a) is not applicable to oilseeds or compound feeds containing milk powder. The sample is extracted with light petroleum and the residue then heated with 3 M HCl. This is filtered, washed, dried and re-extracted with light petroleum.

There are several producers of automatically controlled Soxhlet extractors, which require their proprietary glassware. Some examples are:

Büchi Labortechnik AG: http://www.buchi.com/
Foss: http://www.foss.dk/foss.asp
Gerhardt: http://www.gerhardt.de/gb/soxt.htm
Soxhlet heating mantles are produced by: Electrothermal Engineering Ltd: http://www.electrothermaluk.com/files/prodcore.htm

Fibre, lignin, cellulose, nitrogen-free extract and starch

Fibre can mean many things. Crude fibre is an attempt to measure the roughage material in a feedstuff that is indigestible as far as the animal is concerned. It is an attempt to approximate the effect on the feedstuff of the digestive processes within the digestive tract by the use of inorganic chemicals, in this case, boiling dilute sulphuric acid, then boiling dilute sodium hydroxide, and the weight loss on ignition (which corrects for mineral ash content) of the residue is the fibre content.

There are many modifications of this method, they may be to make the process more representative of the ruminant digestive system, or the desired residue may be just the plant cell walls. It is not always possible to say that one procedure is better than another, therefore the chosen procedure may be that which has been used by workers involved in animal nutrition over a number of years in a certain geographical area. The decision may be to use the usual procedure favoured by the referees for research papers in a particular journal.

For several decades one of the leading authorities on the extraction of fibre from feedstuffs with particular reference to ruminant nutrition has been Professor P.J. Van Soest. His book, *Nutritional Ecology of the Ruminant*, has many helpful details (Van Soest, 1982, 1994). Various detergents are used to fractionate forage matter into its components. Neutral detergent is useful for separating the insoluble plant cell wall fraction, which is only partially digested by ruminal microorganisms (Van Soest and Wine, 1967). He considered that rather than the crude fibre (which entails the loss of soluble fibre components and variable amounts of the hemicelluloses and lignin), it is the proportion of plant cell wall and its degree of lignification that best determines the character and nutritive value of feeds and forage. The cellular contents determine the proportion of completely available nutrients, and consist of the bulk of the protein, starch, sugars, lipids, organic acids and soluble ash. The various processes are given in Table 4.1.

Table 4.1. Effect of detergents and reagents in forage analysis.

Residue type	Reagent	Process	Products
Acid detergent fibre (ADF)[a]	Cetyl trimethylammonium bromide in 0.5 M H_2SO_4	Boil for 1 h	Lignocellulose + insoluble mineral
Neutral detergent fibre (NDF)	Sodium lauryl sulphate, EDTA, pH 7.0	Boil for 1 h	Herbage cell wall minus pectins
Unavailable N	Acid detergent	Kjeldahl nitrogen on ADF residue	Maillard products plus lignified N
Cellulose[a]	None required	Ash from lignin step	By weight loss
Lignin[a]	72% H_2SO_4 on ADF	3 h @ 20°C	Crude lignin
Hemicellulose	Not required	Calculate NDF-ADF	Hemicellulose by difference
Silica (SiO_2)	Conc. HBr treatment of ADF ash	Add dropwise to ash,1 h @ 25°C	SiO_2 residue

[a]Van Soest and Wine (1968).

In the NDF method, the sodium lauryl sulphate (sodium dodecyl sulphate) forms strong protein complexes which are soluble under the right conditions. The EDTA-disodium salt complexes with any Ca or Mg which would otherwise be included with the cell walls. The addition of sodium sulphite to cleave disulphide bridges in any added animal protein (e.g. keratin) in the feed is usually omitted unless major quantities of such substances are present. This is because the sulphite will also dissolve cell wall lignin, reducing its recovery (Moir, 1982; Van Soest *et al.*, 1991). The reagent 2-ethoxyethanol, which aids solution of starches, is toxic and has been replaced by triethylene glycol (Cherney, 2000). Van Soest later recommended omission of anti-foaming decahydronaphthalene (decalin) because it greatly slowed the filtration step (Van Soest, 1973).

The ADF method has tended to replace the crude fibre procedure, especially when further fractionating the feed into lignin and cellulose. However, for an improved correlation between acid detergent fibre and ruminant digestibility, the modified acid detergent fibre (MADF) method of Clancy and Wilson was developed in Ireland (Clancy and Wilson, 1966). Although NIRS is currently the preferred technique to predict OMD (organic matter digestibility) which is then converted to a ME (metabolizable energy) value, this expensive procedure is rarely available to smaller laboratories. Prediction equations were developed for ME from MADF values, and although somewhat less accurate, they may give a working basis for ration formulation. Some examples are given below:

Fresh grass:	ME (MJ kg^{-1} DM)	= 16.20 – 0.0185[MADF] (Givens *et al.*, 1990)
Grass hays:	ME	= 15.86 – 0.0189[MADF] (Moss and Givens, 1990)
Grass silage:	ME	= 15.0 – 0.0140[MADF] (Givens *et al.*, 1989)

This subject is discussed in depth in Chapter 4, 'Feed evaluation and diet formulation' in the Agriculture and Food Research Council (AFRC) advisory manual *Energy and Protein Requirements of Ruminants* (Alderman and Cottrill, 1993), and more recently by Coleman *et al.* (1999).

With non-ruminants, the only fibre determination required is by neutral detergent. Ruminants and other herbivores, which can partially digest fibre, will need the ADF or MADF methods.

Lignin and cellulose

Lignin, like fibre, is a complex substance. Lignins are phenolic polymers that occur in plant cell walls, and they impart, with cellulose, rigidity to stems. There are several molecular building blocks in lignin. When oxidized with nitrobenzene, lignin from angiosperms (grasses, herbs and flowers) yield *p*-hydroxybenzaldehyde, vanillaldehyde (from coniferyl alcohol component) and syringaldehyde. Lignin from gymnosperms (coniferous trees), however, lacks the syringyl group (Harborne, 1984). A typical lignin structural unit is shown in Fig. 4.2.

Estimation of lignin is complicated by the presence of strongly bound proteins. Other contaminants are carbohydrates, chemically bonded cinnamic acids, cutins and tannins. A partial loss of lignin may also occur in the determination, and it is not yet possible to prepare a pure analytical lignin fraction. The relative merits of about 15 procedures are reviewed by Cherney (2000). In the procedure described by Van Soest and Wine (1968), the crucible plus residue from the ADF method is left to stand for 1.5 h in a buffered potassium permanganate solution to dissolve the lignin. The cellulose residue is reacted with demineralizing solution until white, washed successively with 80% ethanol and acetone, dried overnight at 100°C, cooled and weighed. The loss in weight is equal to the lignin content. The crucible may be ashed for 3 h at 500°C and the loss in weight is the cellulose content.

Nitrogen-free extract (Nifext)

This is obtained by subtracting the sum of the percentages of water, protein, fat, fibre and ash from 100. It represents the starch, gums, sugars and organic acids (all N-free), which may be extracted by water or diastase from cleaned, dried and defatted foods. As it is mainly starch, it will be high in the case of cereal grains and lower with seeds containing more oil and protein.

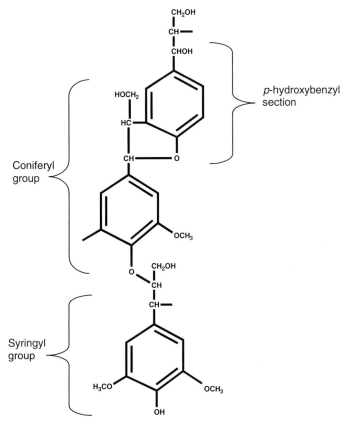

Fig. 4.2. Typical lignin structural unit.

Starch

Starch may be determined using specific enzymes such as amyloglucosidase, but the extraction and hydrolysis stage is slow, enzyme activity can vary, reagents are expensive and complete hydrolysis is difficult. Although acid-hydrolysis lacks specificity, with the simple case of starch in potatoes, it becomes an ideal procedure (Faithfull, 1990). The freeze-dried, milled sample is washed with 10% v/v ethanol/water to remove sugars, dextrins and tannins which can amount to about 12%. *Note:* The use of 80% v/v ethanol/water is recommended for pre-extraction with enzyme methods, followed by heat treatment to gelatinize the starch; 90% v/v ethanol/water tends to make the starch resistant to enzymatic hydrolysis (Hall *et al.*, 2001). The suspension is centrifuged, washed into McCartney bottles using 1 M HCl and heated at 106°C for 40 min. After adjusting the pH to 3.0, it is diluted to 100 ml, and a further dilution with saturated benzoic acid solution provides the solution for analysis of the products of starch hydrolysis. Starch has been given the formula $C_{36}H_{62}O_{31}.12H_2O$, with residue units of $C_6H_{10}O_5$. The hydrolysis is to units of glucose, $C_6H_{12}O_6$, so when using the weight of glucose to determine the initial weight of starch, a correction factor of ×0.9 is required. Although one would anticipate the hydrolysis to yield only glucose, some of the glucose is subsequently converted by the hot acid to fructose and 5-hydroxymethylfurfural which produce more colour with the anthrone reagent than glucose itself. The use of fructose solutions as standards corrects for this effect, otherwise a correction factor of ×0.8 is used with glucose standards. Partial hydrolysis of potato cell walls to chromogenic products led to a further correction factor of ×0.98, and a correction for any moisture must be made.

In vitro digestibility

The estimation of animal digestibility of a feedstuff is usually achieved in one of three ways: *in vivo, in sacco* or *in vitro.* The first uses real animals in feeding trials and gives the most realistic results to which the other methods are correlated; the second method allows feed samples contained in small permeable plastic (e.g. nylon) bags to be inserted through a cannula into the rumen or another section of the digestive tract. The last method allows the digestion of feed samples to occur in the laboratory using digestive juices obtained from a fistulated animal, commercially obtainable enzymes, detergent solutions, or any combination of these, with the aim of imitating naturally occurring digestive processes.

Extractions using detergent plus enzyme

Neutral cellulase plus gamanase digestibility (NCGD) of feeding stuffs. This method originated at a time when compound feeds contained less starch and more digestible fibre and oil than when ME prediction equations were derived in 1985. In this method published by MAFF (1993b), using a fat-free sample, neutral detergent removes soluble cell contents, α-amylase dissolves starch, while cellulase/polysaccharase dissolves cellulose and hydrolyses any polysaccharides in the feed, and gamanase hydrolyses galactomannans which occur in palm kernel products.

Neutral detergent (plus amylase) fibre (NDF) of feeding stuffs. This other MAFF (1993c) method removes cell contents from the fat-free sample by boiling with neutral detergent solution. The α-amylase converts any starch (which would enhance the fibre content) to soluble sugars. The residue is designated neutral detergent (plus amylase) fibre; the abbreviation given is NDF, but this would confuse it with the Van Soest and Wine NDF. Perhaps ND(+A)F would be clearer.

Rumen liquor plus neutral detergent

To obtain the *in vitro* true digestibility, the residue from the first buffered rumen liquor stage of the Tilley and Terry (1963) procedure is digested with neutral detergent solution. The ordinary true digestibility is found by subjecting the faeces to neutral detergent digestion. The neutral detergent soluble non-cell-wall fraction of faeces equates to the endogenous and bacterial loss.

Tilley and Terry (1963) procedure

This eponymous method is widely used and as originally proposed or in modified form has served as a benchmark for other methods. In fact, it is often referred to simply as the *in vitro* digestibility. The first stage involves anaerobic incubation at 38°C in the dark with partially filtered rumen liquor which has been buffered with McDougall's artificial saliva solution, previously saturated with CO_2.

After 48 h, 5 ml of M Na_2CO_3 is added to aid sedimentation immediately before centrifugation. Although mercuric chloride was added to inhibit bacterial activity, immediately centrifuging after 48 h rather than storing samples avoids this. It also avoids disposal of a toxic reagent. The supernatant is decanted into a fine nylon cloth filter and any particles returned to the tube. The particles on the rubber stopper and those adhering to the sides of the tube are washed down to the pellet which is then broken up before adding the acid pepsin solution. This is incubated for a further 48 h, then filtered through a porous alumina crucible (unpublished modification to the original method) before oven-drying, weighing, and possibly ashing.

In vitro *calculations*

The Tilley and Terry method (*X*) correlates with *in vivo* results (*Y*) as follows:

$$Y = 0.99X - 1.01$$

One particular correction is advisable. Standards of known *in vivo* and *in vitro* values covering the lower and higher digestibility range (about 50% and 70% respectively) should be obtained, possibly from a research station, and included with the sample batch. Rumen liquor varies in potency from week to week, therefore a proportional adjustment must be made to enable comparison of results from analyses performed at different times. This variation does not equally affect the low and high standards; one may decrease and the other increase. A graph should be drawn relating the difference of the measured standard from the stated value to the concentration. This could be a positive or negative slope. The samples' measured digestibility should be corrected according to the corresponding adjustment read off the graph. A typical example is shown in Fig. 4.3.

Alternatively, a spreadsheet program such as Microsoft Excel may be used to achieve the correction automatically. A typical example is shown in Table 4.2.

Digestibility equations

There are several ways of expressing the *in vitro* rumen liquor digestibility of a sample: the DOMD or D-value, the DMD value and the OMD value. These are defined below:

1. The DOMD (D-value) is the digestible organic matter in dry matter:

$$= \frac{\text{OM sample} - (\text{OM residue} - \text{OM blank})}{\text{DM sample}} \times 100\%$$

where OM is the organic matter in the original dried and milled sample (sample minus sample ash), OM residue is the organic material in the residue

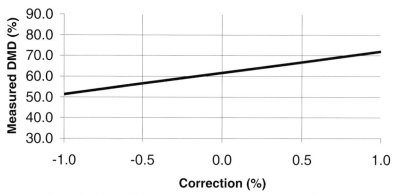

Fig. 4.3. Typical graph for correcting measured sample dry matter digestibility (DMD) values in proportion to deviation of low and high standards from their declared values.

Table 4.2. Typical spreadsheet for correcting the measured sample dry matter digestibility values in proportion to deviation of low and high standards from their declared values.

Spreadsheet for correction of digestibility values between batches			
	Stated value (%)	Measured value (%)	Correction required
High standard	72.9	71.9	1.0
Low standard	50.9	51.4	-0.5
	Corrected value (%)		
Sample 1	51.8	52.2	
Sample 2	62.5	62.2	
Sample 3	73.2	72.2	

Let correction graph be y = mx + c		Spreadsheet formulas	Result
	y = correction to be applied		
	m = slope	((B4-C4) - (B5-C5))/(C4-C5)	0.073
	x = measured sample value	C8	52.200
	mx = Q.10	((B4-C4)-(B5-C5))/(C4-C5)*C8	3.820
	c = intercept on y-axis	(D4-(C4/(C4-C5))*((B4-C4)-B5-C5)))	-4.261
	Vc = corrected value; Vm = measured value		
$y = mx_1 + c =$		((B4-C4)-(B5-C5))/(C4-C5)*C8+(D4-(C4/(C4-C5))*((B4-C4)-(B5-C5)))	-0.441
$y = mx_2 + c$		((B4-C4)-(B5-C5))/(C4-C5)*C9+(D4-(C4/(C4-C5))*((B4-C4)-(B5-C5)))	0.290
$y = mx_3 + c$		((B4-C4)-(B5-C5))/(C4-C5)*C10+(D4-(C4/(C4-C5))*((B4-C4)-(B5-C5)))	1.022
Vc = Vm + y	For Sample 1:	B8 = C8 + y	C8+E20	51.759
	For Sample 2:	B9 = C9 + y	C9+E21	62.490
	For Sample 3:	B10 = C10 + y	C10+E22	73.222

following digestion (residue weight minus ashed residue weight), OM blank is the organic matter in the rumen liquor itself, and DM sample is a dry matter determination done on a separate sample.

This equation must be translated into the actual weighings required so that a spreadsheet can be drawn up. Errors can easily occur in the calculations unless the individual steps are understood. The sample weight is 0.5000 g and the calculation formula and any spreadsheet must be designed for this and allow for the fact that the original ash is carried out on 1.0000 g. The residue of undigested sample contains four components of the calculation:

- undigested sample organic matter
- sample residue ash
- blank (rumen liquor) organic matter
- blank ash.

We are interested in the first component, so need to subtract the other components. The residue from the rumen liquor blank contains both blank organic matter and blank ash. When this value is subtracted from the above we get the sum of sample organic matter plus sample ash.

After weighing the dried residue it is subsequently ashed and weighed. This gives an ash comprising:

- sample residue ash
- blank ash.

A separate ashing of a rumen liquor blank sample gives a figure for blank ash. When subtracted from the above residue ash, the difference gives the sample residue ash. Subtracting this value from the sum of sample organic matter plus ash, leaves us with the undigested residue sample organic matter. Finally, this is subtracted from the original sample organic matter to give the amount of digestible organic matter, which is corrected for dry matter content of the sample and expressed as a percentage or as g kg^{-1} digestibility.

$$DOMD =$$

$$\frac{\left\{\left(\begin{array}{c}0.5 - \text{original} \\ \text{ash}\end{array}\right) - \left(\left(\begin{array}{c}\text{sample residue} - \\ \text{blank residue}\end{array}\right) - \left(\begin{array}{c}\text{sample residue ash} - \\ \text{blank residue ash}\end{array}\right)\right)\right\}}{0.5} \times 100\%$$

2. The OMD value is the <u>o</u>rganic <u>m</u>atter <u>d</u>igestibility:

$$= \frac{\text{OM sample (OM residue OM blank)}}{\text{OM sample}} \times 100\%$$

$$OMD =$$

$$\frac{\left\{\left(\begin{array}{c}0.5 - \text{original} \\ \text{ash}\end{array}\right) - \left(\left(\begin{array}{c}\text{sample residue} - \\ \text{blank residue}\end{array}\right) - \left(\begin{array}{c}\text{sample residue ash} - \\ \text{blank residue ash}\end{array}\right)\right)\right\}}{0.5 - \text{original ash}} \times 100\%$$

3. The DMD value is the <u>d</u>ry <u>m</u>atter <u>d</u>igestibility:

$$= \frac{[\text{sample} \ (\text{sample residue} \ \text{blank residue})]}{\text{sample}} \times 100\%$$

$$\textbf{DMD} =$$
$$\left\{ 0.5 - \left(\text{sample residue} - \text{blank residue} \right) \right\} \times 200\%$$

A suggested spreadsheet for the above calculations is shown in Table 4.3. The values for residues and ash are entered from results sheets printed with columns for crucible weights, etc., unless the laboratory is equipped with computerized balances, when a more sophisticated spreadsheet could be devised.

When planning for Tilley and Terry digestibilities, it is common practice to ensure that the sheep or cattle have been fed for a couple of weeks on a basal diet similar to the test samples to be analysed. This is to ensure a build-up of the appropriate rumen flora resulting in a corresponding optimal activity. Whether or not this is necessary is open to question, and this and other sources of error have been discussed by Ayres (1991). It is also customary not to feed the animal on the morning planned for extracting the rumen liquor.

4. <u>T</u>rue <u>d</u>ry <u>m</u>atter <u>d</u>igestibility (True DMD) (Van Soest *et al.*,1966)
   This is expressed by the equation:

   True DMD = {(% cell content in DM × 0.98) + (% digestible cell wall in DM)}

The % cell content in DM is (100 – % cell wall in DM), which is derived
   from (100 – NDF).
The % digestible cell wall in DM is the (% cell wall in DM – % indigestible
   cell wall in DM).
The % indigestible cell wall in DM is the residual DM after digestion in rumen
   liquor (48 h) followed by the neutral detergent procedure and expressed
   as % sample DM.

Various aspects of *in vitro* methods, from its first use in 1880 to the 1980s have been discussed by the author (Faithfull, 1984). In particular, the effect of pH on tannin complexes, phosphates and sulphides have been studied.

The concept of fistulated animals may seem abhorrent. It should be observed, however, that properly tended animals appear to be quite con-tented, and that their lifetime as an experimental animal is far longer than it would otherwise have been. Nevertheless, it is impossible to prevent the animal from knocking the cannula, and it is easy for leaks to occur causing irritation to the skin around it. It is also expensive to maintain such animals in an acceptable way, and to justify this if long periods exist between experiments. The procedure is favoured by experienced researchers as it facil-itates comparison of results with earlier published work, and may give more consistent results over periods of time. However, improved within-batch precision, economy of time, money and convenience, and improved public

Table 4.3. Typical spreadsheet in Microsoft Excel for calculating the various digestibility values.

Name

In vitro DOMD, OMD & DMD measurement — Corrected

Sample ID	Sample wt.	Sample DM	Fractional sample DM	S dry wt.	Sample ash %.	Fractional ash	Sample OM	Residual wt.	Residual ash	Residual OM	OMD g/kg	DOMD g/kg DM	DMD g/kg	Average
1	0.5000	98.01	0.9801	0.4901	7.24	0.0355	0.4546	0.1891	0.0085	0.1775	610	565	614	=AVERAGE
2	0.5000	97.22	Formula =D5/100	=samplewt *E5	6.80	=G5/ 100*F5	=F5-G5	0.1773	0.0072	=J5-K5 -M19	=(I5-L5) /I5*1000	=(I5-L5) /F5*1000	=(F5-J5) /F5*1000	(M4:M5)
3	0.5000	98.25	0.9825	0.4913	8.43	0.0414	0.4498	0.1896	0.0088	0.1777	605	554	614	
4	0.5000	97.56	0.9756	0.4878	7.48	0.0365	0.4513	0.1907	0.0079	0.1797	602	557	609	603
5	0.5000	98.18	0.9818	0.4909	6.61	0.0324	0.4585	0.1814	0.0076	0.1707	628	586	630	
6	0.5000	97.73	0.9773	0.4887	7.53	0.0368	0.4519	0.1882	0.0081	0.1770	608	562	615	618
7	0.5000	97.66	0.9766	0.4883	7.96	0.0389	0.4494	0.1901	0.0077	0.1793	601	553	611	
8	0.5000	97.94	0.9794	0.4897	8.01	0.0392	0.4505	0.1866	0.0083	0.1752	611	562	619	606
9	0.5000	98.12	0.9812	0.4906	8.22	0.0403	0.4503	0.1871	0.0080	0.1760	609	559	619	
10	0.5000	96.97	0.9697	0.4849	7.87	0.0382	0.4467	0.1888	0.0077	0.1780	602	554	611	605
Standard L1	0.5000	98.44	0.9844	0.4922	6.98	0.0344	0.4578	0.1903	0.0082	0.1790	609	567	613	
L2	0.5000	97.89	0.9789	0.4895	7.34	0.0359	0.4535	0.1855	0.0090	0.1734	618	572	621	613
H1	0.5000	98.35	0.9835	0.4918	7.97	0.0392	0.4526	0.1899	0.0083	0.1785	606	557	614	
H2	0.5000	97.17	0.9717	0.4859	6.84	0.0332	0.4526	0.1977	0.0076	0.1870	587	547	593	596
M19 1								0.0040	0.0009	0.0031	0.0031			
2								0.0040	0.0009	=J19-K19				

The row for Sample 2 is used to display the formulae, which are normally hidden.

perception all point to alternative methods as being the way forward. One such method uses faecal liquor and has been discussed by Omed *et al.* (2000). Replacing the acid pepsin stage with biological washing liquid produced digestibilities very close to the known *in vivo* values for a variety of grasses, legumes and hays (Solangi, 1997). The two-stage pepsin–cellulase method (see below) is probably the best alternative to the Tilley and Terry procedure.

Cellulase digestibility

A convenient procedure for assessing the digestibility of forages is the cellulase digestibility technique. This was refined by Jones and Hayward (1973) at the Welsh Plant Breeding Station (WPBS) in Aberystwyth (since 1992, the Institute for Grassland and Environmental Research). It was later extended to a two-stage procedure with a pepsin pre-treatment (Jones and Hayward, 1975).

The pepsin removes protein from the cell walls and possibly modifies the cell wall polysaccharide in such a way as to render it more susceptible to attack by the cellulase enzyme. It also allows cellulases from different sources to be used with less effect from variation in enzyme activity. The single stage cellulase technique is suggested for screening in plant breeding programmes, but in this case, the higher activity enzyme from *Trichoderma viride* will yield a higher correlation with *in vivo* and *in vitro* digestibility. One might expect less precision when digesting with enzymes versus rumen liquor, because enzymes lack the ability of microorganisms in adapting to a substrate. Stakelum *et al.* (1988), however, found a similar accuracy in predicting *in vivo* digestibility when using the rumen liquor–pepsin, pepsin–cellulase or neutral detergent–cellulase methods.

Nitrate and water-soluble carbohydrate

The same extractant is used for both nitrate and water-soluble carbohydrate (WSC) determinations, however the ratio of sample to extractant is different. The herbage may be oven-dried for nitrate, but must be freeze-dried for WSC determination. The extractant is saturated benzoic acid solution. Benzoic acid is sparingly soluble in cold water, and the solution is made by adding an excess quantity to deionized water at ambient temperature in a blender, which is then switched on for about a minute. It is filtered through a Whatman No. 4 paper into a storage container fitted with a tap. If the ambient temperature should fall several degrees, it is possible for some crystals to separate out. These would make little quantitative difference, but might block the sample capillary probe or tubing. If this is thought likely, the containers for samples and standards should be warmed and shaken gently to redissolve the crystals. The benzoic acid acts as a preservative, allowing the storage of sample extracts at room temperature almost indefinitely, so they can be analysed at a convenient time. It has been noticed, however, that the concentration of nitrite (as opposed to nitrate, which is stable) decreases to zero after a day or so. Methods estimating nitrite, therefore, must use an extractant such as water, followed by immediate analysis.

Nitrate

The autoanalysis method was developed at the WPBS and modified by using benzoic acid extractant solution. It is based on the method of Follett and Ratcliff (1963), which was itself based on that of Grace and Mirna (1957). It relies on the reduction of nitrate to nitrite by adding the sample solution to an ammonium chloride buffer (pH 7.5) containing EDTA disodium salt and copper sulphate and passing through a glass tube containing cadmium filings which become copper-plated. The nitrite immediately reacts with sulphanilamide to form a diazo salt which couples with 8-aminonaphthalene-2-sulphonic acid (Cleve's acid) to form an orange acid azo dye which is measured at 470 nm on a spectrophotometer.

Another method used for nitrate determination on dried and milled herbage employs the nitrate selective electrode. One of the first published methods was that of Paul and Carlson (1968). Other anions, especially chloride, can interfere. These authors removed chloride with silver resin, but Barker *et al.* (1971) omitted the resin because it tended to foul the electrode and cause excessive drift. Normally the $Cl^-:NO_3^-$ ratio is so low as not to interfere, but saline precipitation from coastal plots could affect this. The method was further modified to allow storage of extracts for up to 64 h by adding a preservative of phenyl-mercuric acetate and dioxane, both very toxic (Baker and Smith, 1969). This paper mentions the need to change the electrode's membrane, filling solution and liquid ion exchanger every 2 months to minimize chloride interference. It is easy to overlook electrode maintenance between batches of nitrate analyses, and this can lead to errors and sluggish performance.

The method was extended from plants to include soils and waters by Milham *et al.* (1970). They point out that nitrate reductase activity in fresh plant samples often causes a rapid decline in nitrate content, so samples collected from remote sites should be frozen in dry ice. A trace of chloroform was used to protect soil and water samples before freezing. We are now more aware of the harmful effects of chloroform inhalation and suggest immediate freezing without preservative and analysis within a few days as a safer alternative – especially with student projects.

One drawback with selective ion electrodes is their slow response at low concentrations of analyte, perhaps below 2 mg l^{-1} NO_3-N. It can take several minutes to equilibrate, and slow drifting can give a measure of uncertainty as to the equilibration point. If this cannot be remedied by reducing the dilution factor, an alternative method should be sought. They are also sensitive to changes in temperature, in excess of 1°C being significant. Mechanically driven magnetic stirrers get warm, therefore electronic ones are preferable.

Water soluble carbohydrate

This is basically the anthrone method of Yemm and Willis (1954) which was developed at the WPBS for use with an autoanalyser (Thomas, 1977) and modified by using benzoic acid extractant solution. The extract is reacted with anthrone in 76% sulphuric acid. Heating to 95°C develops the green colour which is measured at 620 nm. Fructose, sucrose and inulin give the colour at room temperature, but heating is necessary for glucose, maltose, fucose and rhamnose to react (Van Handel, 1967). Fructose and glucose are hydrolysed by hot sulphuric acid to 5-hydroxymethyl-furfural which reacts with anthrone to give 10-{5-(anthron-10-ylmethyl)-2-furfurylidene} anthrone, which couples with brown resin by-products to give the colour (Hoermann, 1968).

Water content in silage

Silage moisture consists of both water and volatile fatty acids (VFAs). To oven-dry silage would remove both the water and the nutritional VFAs which should be included with the DM. The most widely used method to correct for this loss is the toluene distillation method of Dewar and McDonald (1961). The Karl Fischer titration is probably the most accurate, but uses anhydrous methanol. Oven drying and using correction equations makes assumptions which may not be valid in every case. NIRS involves very expensive equipment and extensive calibration. Various pros and cons have been discussed by Givens *et al.* (2000). The main deficiency of the toluene distillation is its inability to account for the alcohol content of the volatiles. The other drawback is the large quantities of toluene involved. It is, however, a simple method and is widely quoted. We suggest a smaller scale procedure which uses less solvent, and recovers used solvent by distillation and drying over anhydrous sodium sulphate. If the accuracy requires the alcohol content to be determined, this may be done separately by GLC. The method also enables small core samples to be analysed, which simplifies the profiling of silage clamps for nutrient analyses (Faithfull, 1998).

Extraction Procedures – Soils

There are many different types of soil, and extractant formulations have been fine-tuned to suit the soil. The particular extractant may also be chosen on the basis of familiarity over the years, and because it is easier to compare results with those previously obtained, and hence make recommendations to correct deficiencies based on experience. Usually one is not interested in the total amount of a soil nutrient, rather in the amount that is in a form available to the roots of the plant. Regional advisory laboratories over a long period may have developed index tables relating to the found concentration of nutrient in local soil types and the corrective amount of fertilizer required. It would probably be wise to adopt the same methods that have been used to derive these tables, unless they have been found to be inadequate.

We will refer to the UK MAFF/ADAS publications in the appropriate chapter. There are, however, published procedures on the web, particularly from the USA. Two such manuals are available from Delaware Cooperative Extension (1995): Recommended Soil Testing Procedures for the Northeastern United States, 2nd Edition, and from the Missouri Agricultural Experiment Station (1998): Recommended Chemical Soil Test Procedures for the North Central Region at their respective websites:

http://bluehen.ags.udel.edu/deces/prod_agric/title-95.htm
http://muextension.missouri.edu/xplorpdf/miscpubs/sb1001.pdf

The United States Department of Agriculture (USDA, 1996) has also published a methods manual.

pH extractants

The apparently simplest of procedures faces one with a choice of about four extractants. The commonest extractant is water, and the ratio we use is 10 ml soil:25 ml water, i.e. 1:2.5 v/v. Other ratios used by the Northeastern United States are 1:1 v/v, 1:1 w/v and 1:2 v/v soil/water (Delaware Cooperative Extension, 1995, Appendix). Some soils have a significant soluble salt content, which can affect the measured pH. The concentration of these salts in the soil varies with the season, with dry season pH values being lower than wet season ones. This is because salts such as sulphates and nitrates, which lower pH, accumulate in dry periods and are leached away in rainy periods. To overcome this effect, a 1 M KCl extractant was first used. The pH values so obtained are 1.5–2.0 units less than those with water extractant, and are also affected by variations in the soil:extractant ratio. It is still used to assess the aluminium status of the soil. Values below pH 5 indicate significant amounts of Al, and if very much lower than 5, almost all the acidity is in the form of Al (USDA, 1996, p. 149). The aluminium acts by displacing hydrogen ions from the exchange sites on the surface of clay and humus particles to increase the acidity by raising the hydrogen ion (H^+) concentration.

It was later proposed that more suitable extractants to overcome, and also measure, the salt effects which displace hydrogen ions in a seasonal manner would be either 0.1 M KCl or 0.01 M $CaCl_2$, with the latter being more widely used (Schofield and Taylor, 1955).

The effect on mineral soils with a permanent negative charge, or on organic soils with a negative charge which varies with pH, is to displace H^+ and lower the measured pH by about 0.5 units compared with water extractant. The effect on mineral soils dominated by sesquioxides, kaolinite and allophane with variable charge is that the salt causes adsorption of H^+ onto reactive sites, raising the pH by about 0.5 units (Rowell, 1994, p. 161). The difference in pH between water and salt solution extracts is known as the *salt effect*, and given the symbol Δ pH. Thus,

Δ pH = soil pH in salt solution − soil pH in water

and Δ pH values are positive for soils with a net positive charge, and negative for soils with a net negative charge, with magnitude proportional to charge.

Phosphate extractants

Phosphorus occurs in various soil fractions: as soil minerals combined with Ca, Fe, Al, which are of low solubility; bound to particle surfaces of, e.g. sesquioxides, calcite, to Al on humus surfaces; in soil solution; in the organic matter, primarily as esters.

Again, there are several choices of extractant, and the preferred one depends mainly on the type of soil under test. One of the most widely used procedures is the Olsen method (Olsen *et al.*, 1954), which was developed in the USA to correlate crop response to fertilizer on calcareous soils. The amount of P extracted will vary with temperature (increases by 0.43 mg P kg^{-1} per degree rise between 20°C and 30°C) and shaking speed, so conditions should be standardized. The extractant is 0.5 M sodium bicarbonate adjusted to pH 8.5. The bicarbonate competes with phosphate on the adsorption sites extracts, and removes most, but not all of it, together with some soluble calcium phosphate. Addition of phosphate-free activated carbon before shaking is necessary if coloured soil extracts are obtained, and then they will require filtration.

The northeastern United States have soils where the P chemistry is affected by aluminium phosphates. They therefore use dilute acid extractants to dissolve these minerals and extract the P. They use several procedures: (i) The Mehlich 1 Extraction (dilute double acid extractant) containing 0.0125 M H_2SO_4 + 0.05 M HCl (Mehlich, 1953). (ii) The Mehlich 3 Extraction using 0.2 M acetic acid + 0.25 M ammonium nitrate + 0.015 M NH_4F + 0.013 M nitric acid + 0.001 M EDTA (ethylenediaminetetraacetic acid) (Mehlich, 1984). The pH should be 2.5. (iii) The Morgan Extraction using 0.72 M NaOAc (sodium acetate) + 0.52 M acetic acid at pH 4.8 (Morgan, 1941). (iv) The Modified Morgan Extraction (McIntosh, 1969) using 0.62 M NH_4OH + 1.25 M acetic acid at pH 4.8. The resulting extracts are used for the appropriate colorimetric reaction and absorbances are measured on a colorimeter or spectrophotometer, possibly coupled to an autoanalyser.

Note: the Olsen method is not to be confused with the Olson method (Olson *et al.*, 1954), which uses sodium carbonate.

The North Central Region, in addition to the Olsen method, uses the Bray and Kurtz P-1 test for phosphorus (Bray and Kurtz, 1945), which has proved to be well correlated with crop response to phosphate fertilizer on acid to neutral soils in the region. Each state experiment station has developed correlations and calibrations for the particular soil conditions within its own state, so field experience over a number of years or decades is necessary when deciding which methods to adopt. When bringing samples from remote sites back to the laboratory, it is therefore important to assess the nature of the soil at that site in order to choose the optimum method. If the same method has to be used for reasons of comparability, then it is necessary to state that the available phosphorus content was obtained using a particular named method.

Another method used by a few laboratories will be briefly mentioned; that is determination by resin extraction. The latest fertilizer recommendations by MAFF/ADAS (2000) include a classification of soils from the resin P values obtained using the method of Hislop and Cooke (1968). This was developed in the 1960s at the Levington Research Station, Ipswich. The method was intended to reflect the soil phosphate capacity, intensity and kinetic (rate of release) components. It was also designed to avoid inducing any major change in the chemical constitution of the soil as a result of the applied extraction procedure. The anion exchange resin (De Acidite FF 510, particle size >0.5 mm) was considered to be an inert phosphate sink. The method is outlined as follows: a subsample of 20 g air-dry soil, ≤2.0 mm, is ground in a Glen Creston Micro Hammer Mill fitted with a 0.5-mm screen. A 2-ml scoop of soil is then transferred to a 6 ounce (170 ml) bottle followed by a 5-ml scoop of washed and dried resin and 100 ml distilled water. It is shaken end-over-end for 16 h at 25°C, after which it is filtered through approximately 0.5 mm terylene netting and washed; this retains the resin and allows the soil to pass through. The resin is then transferred to a leaching tube and 50 ml sodium sulphate (70 g l^{-1}) solution is added and the leaching controlled to last 20 min. The phosphate in the leachate is determined colorimetrically, either manually or using an autoanalyser, the method being based on Fogg and Wilkinson (1958). The resin procedure was correlated with the Olsen bicarbonate method and gave a correlation coefficient of 0.877 (significant at $P \leq 0.001$) for non-calcareous soils, and 0.830 for calcareous soils. The amount of phosphate extracted by the resin during 16 h shaking approaches a maximum, and reflects the quantity or capacity factor which dominates under agricultural conditions. For glasshouse soils, however, full extraction is not approached, and it is rather the intensity and kinetic factors which are reflected more than capacity, and which are considered to be more relevant in this situation.

Others use the resin method of Somasire and Edwards, 1992. The latter involves extracting 5 g soil 1:20 (m/v) using 100 ml of water, 2.8 ml cation exchange resin and 4.0 ml anion exchange resin with shaking for 16 h; this is followed by extraction with 1 M ammonium chloride, pH 2.0, with 30 min shaking.

The above extractants for phosphate have been mainly developed for conventional agriculture. Some methods have been developed for assessment of soils managed on the organic system, which will be discussed in a separate chapter.

Tip: finding soil analysis methods on the web requires a more powerful search engine. Try searching for 'soil test procedures' using http://www.alltheweb.com or http://www.google.co.uk

Potassium extractants

Potassium occurs in soil clay minerals, feldspars and micas. The unweathered *illite* region of the clay mineral contains non-exchangeable K^+, the weathered *vermiculite* region has exchangeable K^+, while the intermediate region has slowly exchangeable K^+. There is also available potassium in the soil solution. The extractant will leach the free potassium ions and displace the exchangeable and some slowly exchangeable K^+ by replacing the K^+ with Na^+, H^+ or NH_4^+, depending on the extractant. Various extractants are listed in Table 4.4.

Table 4.4. Some extractants for potassium in soils used by various regional laboratories.

Regional methodology	Extractant	Comments
North Central USA	Mehlich 3	Non-calcareous soils
	1 M ammonium acetate	Calcareous soils if Ca and Mg also to be extracted
North Eastern USA	Morgan, Modified Morgan, Mehlich 1 and Mehlich 3	All are acidic and could extract some non-exchange-able K^+. Ammonium based reagents extract more K^+ than H^+ or Na^+ based ones
MAFF/ADAS UK	1 M ammonium nitrate	Also used for Mg and Na

Trace element extractants

The determination of total amounts in soil is valid for finding whether there are toxic levels of certain metals (e.g. after repeated slurry applications), and comparisons can be made with published tables of maximum recommended levels. Some typical and maximum values are shown in Table 4.5 (ADAS, 1987; DOE/NWC, 1981). Dutch values differ from those developed in the UK in that the intention is to allow the return of contaminated land to any potential use, rather than tailoring the level of remediation to the intended use of the land. The most recent values include general targets and intervention values (http://www.athene.freeserve.co.uk/sanaterre/guidelines/dutch.htm).

The soil sample is ground to pass a 0.5-mm sieve, and 2.5 g taken for the analysis. There are two possible extractants. Firstly 1:4 $HClO_4$ (60% by weight perchloric acid):HNO_3 (70% by weight nitric acid), of which 25 ml is added to the sample. It is allowed to stand overnight, then heated at 100°C, next 180–200°C and finally at 240°C. The residue is dissolved in 6 N HCl, boiled, cooled, made to 50 ml and filtered before analysis by atomic absorption spectrophotometry. This can be a dangerous procedure with a risk of explosion, and the full details should be carefully followed as given in the original reference (MAFF/ADAS, 1986, p. 31). An alternative acid mixture,

Table 4.5. Typical and maximum recommended levels for some trace elements in soil.

Metal	Typical value in uncontaminated soil		Maximum recommended level		Earth/sediment (mg kg⁻¹ dry matter)	
	(mg kg⁻¹)	(kg ha⁻¹)[a]	(mg kg⁻¹)	(kg ha⁻¹)[a]	Target value	Intervention value
Zinc	80	160	300	600	140	720
Copper	20	40	135	270	36	190
Nickel	25	50	75	150	35	210
Cadmium	0.5	1	3	6	0.8	12
Lead	50	100	250	500	85	530

[a]Assumes 2000 t ha⁻¹ to depth of 15 cm.

aqua regia, is now suggested, and although a highly corrosive reagent, there should be no risk of an explosion.

The availability of the trace metals is easily determined without any of the above risks, and the results used to assess both deficiencies and toxicities. The metals need to be removed from the sites where they are bound to the soil particles by use of an even stronger binding agent than the soil. This is achieved with two possible complexing reagents: EDTA and DTPA. They are a class of chemicals known as complexones, which form complex molecules with metals in a cage-like structure called a chelate.

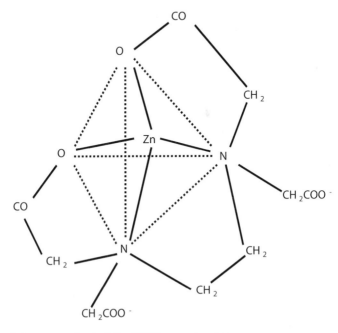

Fig. 4.4. A four co-ordinated Zn-EDTA complex ion.

EDTA is ethylenediaminetetraacetic acid, also called (ethylenedinitrilo) tetraacetic acid, mol. wt 292.24. Although having four carboxylic groups, it behaves as a dicarboxylic acid with two strongly acidic groups. It is used as the disodium or ammonium salt, the latter being formed *in situ*. One mole of the EDTA salt reacts in all cases with one mole of the metal irrespective of its valency state. A four co-ordinated zinc EDTA complex is shown in Fig. 4.4, and a six co-ordinated cobalt EDTA complex in Fig. 4.5. EDTA is known as hexa- (or sexa-) dentate, having up to six active metal-complexing sites per molecule.

DTPA is diethylenetriaminepentaacetic acid, also known as diethylenetrinitrilopentaacetic acid, mol. wt 393.36. It is octo-dentate, having eight active metal-complexing sites per molecule. A diagrammatic representation of the DTPA molecule is shown in Fig. 4.6.

The amount of metal extracted from the soil by both EDTA and DTPA is dependent on the pH, the metal being extracted, the soil:solution ratio, the concentration of chelating agent, the shaking time, the temperature, and the sample preparation procedure. Clearly, the methodology used should be clearly described and closely followed if repeatable work is to be possible, and comparison of results is to be meaningful.

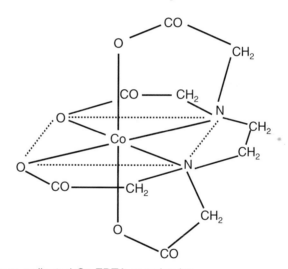

Fig. 4.5. A six co-ordinated Co-EDTA complex ion.

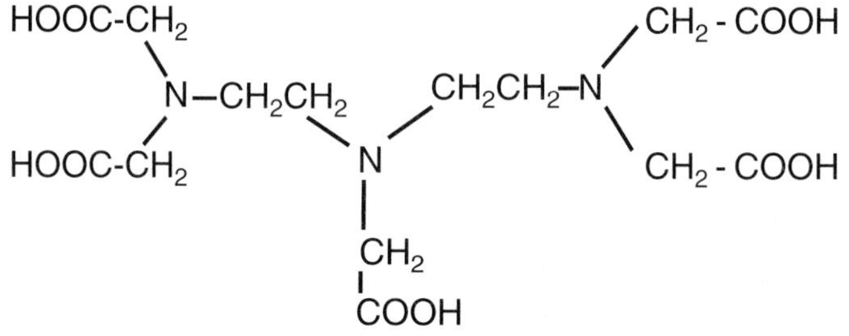

Fig. 4.6. A diagrammatic representation of a molecule of DTPA.

5 Analysis of Soil and Compost

Soil Analytical Procedures

Method 5.1. Determination of extractable boron

The predominant form of boron in soil solution is H_3BO_3, but above pH 9.2, H_2BO_3 may predominate. Hot-water extraction is the most widely accepted procedure for determining the amount of boron that is available to plants, and correlated best with the incidence of black spot in garden beets (Missouri Agricultural Experiment Station, 1998). The final determination is best performed using an ICP spectrometer, but this may not always be available, so a colorimetric method will be described. Methods using either curcumin or azomethine-H are possible, but the latter will be suggested here. It is not only the reagent used in the MAFF/ADAS (1986, pp. 20–22) handbook, which is the method to be described (with Crown Copyright permission), but has been adopted by the Delaware Cooperative Extension (1995) as being rapid, reliable and requiring less sample preparation and handling than the curcumin method. The American methodology, however, omits the removal by ashing of any organic matter in the filtrate, which might interfere with the determination; it also adds 0.1% m/v $CaCl_2.2H_2O$ to the water extractant to promote soil flocculation. Also, the filtration step is replaced by centrifugation in a plastic centrifuge tube at 2700 g for 15 min. The following method could be modified similarly if appropriate, but once adopted, should be adhered to for future comparison of results.

Boron is obviously a component of borosilicate glassware, which should therefore be avoided. Apparatus should therefore be made of PTFE, soda glass or silica. A fibre digestion apparatus may be suitable. It may even be possible to extract the soil by boiling with the extracting water in a sealed plastic bag or pouch (Mahler *et al.*, 1984). Having said that, however, silica (or quartz) apparatus is expensive, especially for educational purposes. The comment by Bingham (1982) should therefore be noted: 'We have not found it necessary to use special low-B glassware for the analysis of water, soil, or plant samples. Pyrex glassware or plastic ware has been entirely satisfactory.' Presumably the magnitude of the blank reading would show whether there was a contamination by extraneous boron.

Apparatus.

- Flasks, 250 ml, conical with ground joint – silica (quartz), or soda glass.
- Condenser – either silica cold finger condensers, effective length 140 mm, or soda glass air condensers, approximately 750 mm.
- Evaporating basins – 20 ml, translucent silica, shallow form, with round bottom and spout.
- Polyethylene tubes – 20 ml with hinged cap.

Reagents. Note: all reagents must be stored in polyethylene containers.

- Azomethine-H reagent – Dissolve 0.45 g of azomethine-H in 100 ml of 1% m/v L-ascorbic acid solution. Prepare fresh weekly and store in a refrigerator.
- Boron stock standard solution, 100 µg B ml^{-1} – Dissolve 0.572 g of boric acid (H_3BO_3) in water and dilute to 1 l and mix.
- Boron intermediate standard solution, 20 µg B ml^{-1} – Pipette 20 ml of the boron stock standard solution into a 100 ml volumetric flask and make up to the mark with water and mix.
- Boron working standard solutions, 0–3 µg B ml^{-1} – Pipette 0, 1.0, 2.0, 5.0, 10.0 and 15.0 ml into 100 ml volumetric flasks and make up to the mark with water and mix. This will provide solutions containing 0, 0.2, 0.4, 1.0, 2.0 and 3.0 µg ml^{-1} of boron.
- Buffer masking reagent – Dissolve 250 g of ammonium acetate and 15 g of EDTA, disodium salt, in 400 ml water. Carefully add 125 ml of glacial acetic acid.
- Calcium hydroxide solution, saturated.
- Hydrochloric acid, approximately M – Dilute 85 ml of hydrochloric acid, approximately 36% m/m HCl, to 1 l with water.
- Sucrose.

Procedure. Transfer 40 ml (2 × 20 ml plastic scoopfuls, struck off level without tapping) of air-dry soil, sieved to ≤2 mm, into a flask. Measure 80 ml of cold water into a boron-free container and bring to the boil. Transfer the boiling water to the flask containing the soil and attach a condenser. Reheat to boiling as quickly as possible, and continue to boil for exactly 5 min. Remove the flask from the heat source, and allow to stand for exactly 5 min. Filter under reduced pressure through a 125 mm Hartley funnel fitted with a 125 mm

Whatman No. 2 filter paper, collecting the filtrate in a boron-free tube inside the filter flask. Terminate the filtration after 5 min, and retain the filtered extract for the determination of boron. Carry out a blank determination.

Pipette 5 ml of each boron working standard solution into a silica evaporating basin, and add 0.25 g of sucrose and 2 ml of satd calcium hydroxide solution. Evaporate to dryness on a boiling water bath. Place the basin in a cold muffle furnace, slowly increase the temperature to 450°C and maintain this temperature for 2 h. Allow to cool, then add exactly 5 ml of approximately M hydrochloric acid and dissolve all soluble material. Filter through a 90-mm Whatman No. 541 filter paper. Transfer 1 ml of the filtrate to a polyethylene tube. Add 2 ml of buffer masking reagent, mix and add 2 ml of azomethine-H reagent. Mix well and allow to stand for 45 min. Measure the absorbance in a 10 mm optical cell at 420 nm. Construct a graph relating absorbance to μg of boron present. The absorbances corresponding to 0 and 3.0 μg of boron are approximately 0.2 and approximately 0.7 and should differ by approximately 0.45.

Pipette 5 ml of each soil extract into a silica evaporating basin, add 0.25 g sucrose and 2 ml of satd calcium hydroxide solution, and proceed as above as far as measuring the absorbance at 420 nm.

Calculation. Read from the standard graph the number of μg boron equivalent to the absorbance of the sample, and the absorbance of the blank. Multiply the difference by 2 to give the mg l^{-1} of boron in the air-dry soil sample. To express in terms of oven-dry soil, see Method 5.2, Calculation (2).

Method 5.2. Cation exchange capacity, exchangeable bases and base saturation

Definitions

Percentage base saturation:	$= 100 \times TEB/CEC_7$
Cation exchange capacity (CEC):	the sum total of exchangeable cations that a soil can adsorb
CEC_7 or CEC-7:	the CEC determined with 1 M ammonium ethanoate (ammonium acetate) buffered at pH 7.0
Effective cation exchange capacity (ECEC):	
	the sum of the exchangeable cations (Al^{3+}, H^+, Ca^{2+} and Mg^{2+}) extracted by 1 M potassium chloride
Total exchangeable bases (TEB):	the sum of the exchangeable 'basic' cations (Ca^{2+}, Mg^{2+}, K^+, Na^+ and NH_4^+) acted with 1 M ammonium ethanoate at pH 7.0.

Discussion. The colloidal clay and humus soil fractions are negatively charged and therefore attract and adsorb positive ions (cations) on to exchange sites. These may be the so-called basic cations defined above, or the acidic cations H^+ and Al^{3+}. These cations are not soluble in water when in the adsorbed state, but can exchange with H^+ which is present in the acidic vicinity of the plant root system. They are now in solution and able to be absorbed into the plant. The extent to which the exchange sites are saturated with cations, together with the ratios of the cations to each other, indicates the nutrient supplying power of the soil.

The principle behind the determination of the CEC is that ammonium ions will leach the adsorbed metallic cations from the soil (soil ammonium having a ratio small enough to be ignored in this group of calculations) as a solution suitable for analysis by flame emission and atomic absorption techniques. The reagent M ammonium ethanoate is universally adopted for this purpose. The presence of any free basic cations as salts in solution would give an exaggerated TEB value, therefore some workers suggest an initial leaching with aqueous ethanol. This may be 95% ethanol, or more economically for class work, 95% or even 60% industrial methylated spirits (IMS), which is also used to remove excess ammonium ethanoate. The initial leaching is not usually necessary for temperate (UK) soils. In certain cases, ethanol may remove some adsorbed NH_4^+, and should be replaced with isopropanol. The amount of ammonium ion adsorbed on to all the exchange sites is a measure of the CEC. It may be determined either by leaching with acidified KCl (100 g l^{-1} KCl + 2.5 ml M HCl) to remove the ammonium ions, then a 25 ml aliquot of this solution is made alkaline to convert NH_4^+ to NH_3 which is steam-distilled over and titrated, or the entire soil sample may be steam-distilled. The latter method has two disadvantages: it is difficult to transfer the entire soil sample to the distillation flask, and any non-exchangeable ammonium in the sample could be liberated to give an inflated CEC value.

The TEB value may be obtained by either the sum of the individually measured cations or by evaporating and igniting a portion of the ammonium ethanoate leachate to convert the metallic cations to oxides and carbonates, followed by addition of excess acid (to convert carbonates to chlorides) and back-titration with alkali. The latter method is difficult if the soil is insufficiently base-rich to provide an adequate amount of bases for the titration. On the other hand, the calcium carbonate in calcareous soils may be partially leached by the ammonium ethanoate at pH 7.0 in addition to the exchangeable bases and thus give an exaggerated TEB value and a percentage base saturation in excess of 100%. The TEB by ignition/titration can serve as a check on the values from the summation method.

If the percentage base saturation as defined above is ⩽60%, this provides an indication of the need for estimation of exchangeable aluminium and hydrogen, in addition to calcium and magnesium, by the ECEC procedure.

It must be strongly emphasized that the charge on the humus and mineral particles depends not only on the nature of the surface but on the pH, the negative charge, hence CEC, rising with increase in pH. The CEC_7 can therefore be far higher than it would be in the field. It is therefore necessary to ascertain the experimental conditions when assessing published data. One way to compensate for this effect is to carry out the leaching with unbuffered 1 M KCl solution (some methods use unbuffered NH_4Cl), which will not affect the *in situ* pH of the soil. Obviously this precludes subsequent analysis of potassium, but this is one of the minor cations. This analysis is termed the effective cation exchange capacity (ECEC). The exchange complex in acidic soils tends to be dominated by Al^{3+} rather than H^+ according to the reaction:

$$3H^+ + Al(OH)_3.3H_2O \rightleftharpoons Al^{3+} + 6H_2O$$

The solubilized aluminium is the main toxic agent to plants in acidic soils, and acid tolerant (calcifuge) plants are usually also aluminium tolerant. The ECEC method determines the levels of Al^{3+}, H^+, Ca^{2+} and Mg^{2+} extracted by 1 M potassium chloride and is described in Method 5.2.

A detailed discussion of the above topics together with a selection of class projects and test calculations is given in Chapter 7 of Rowell (1994, pp. 131–152), but note his calculation of CEC has an error in that it is based on 250 ml KCl extract, not on 100 ml as per given methodology, and thus requires correction (confirmed by personal communication, 2001). Directions for using the Foss/Perstorp Analytical Tecator Kjeltec Auto 1035/1038 Sample System for CEC determinations are given in USDA, 1996 (pp. 203–210).

Determination of CEC and exchangeable cations

Reagents. Note: deionized water and analytical grade chemicals are used throughout unless otherwise stated.

- Ammonium ethanoate, M – dilute approximately 230 ml glacial ethanoic (acetic) acid to 1 l. Dilute approximately 220 ml ammonia solution (ammonium hydroxide) approximately 35% m/m NH_3 to 1 l in a fume cupboard. Mix together in a 5-l graduated beaker and adjust the pH to 7.0 using ethanoic acid or ammonia solution added using a disposable polyethylene pasteur pipette. Stir with a glass rod between additions, but allow solution to become still before reading the pH. Dilute to 4 l and transfer to a polythene storage bottle.
- Ethanol, 95% (or industrial methylated spirits, 95%) – dilute ethanol (or IMS) to give 95% v/v ethanol/water.
- Potassium chloride solution – dissolve approximately 100 g KCl in water and make up to 1 l. Add 2.5 ml M HCl, and check the pH is approximately 2.5.

Extraction. Transfer 5 g sieved ($\leqslant$ 2 mm) air-dried soil to a 100-ml glass beaker, add 20 ml M ammonium ethanoate, stir and let stand overnight. Transfer the contents to a filter funnel fitted with a 125 mm Whatman No. 44 filter paper and held in a 250 ml volumetric (graduated) flask. Wash the beaker with ammonium ethanoate reagent from a wash (squeeze) bottle to remove all the sample, then add successive 25 ml volumes of reagent to leach the soil in the funnel, allowing it to drain between additions. With the collected leachate volume approaching 250 ml, remove the funnel to a rack or place in a 250 or 500 ml conical flask, and make up the volume to the 250 ml mark with reagent and retain for analysis of exchangeable bases.

The soil in the funnel is washed free of excess reagent by five successive additions of 95% ethanol, allowing to drain between washings. A wash bottle containing ethanol enables the interior surface of the funnel, the outside of the stem, the exposed surface of the paper and the soil to be thoroughly washed. Any remaining ammonium ethanoate will elevate the final CEC value. The washings, which are flammable, should be collected in a waste solvents bottle for safe disposal.

The funnel is now placed in a 100-ml volumetric flask and leached with successive 25-ml portions of potassium chloride solution, allowing draining between additions, until nearly 100 ml has been collected. Make up to the mark and retain for determination of CEC.

Measurement of calcium and magnesium by AAS

This is achieved by using an atomic absorption spectrophotometer (or less accurately with a flame photometer). Some details could be instrument specific, so refer to the manufacturer's handbook, application data sheets, and obtain technical support if you lack experience in this area. Some general guidelines will be noted here.

The use of a nitrous oxide-acetylene flame obviates the need for releasing agents to be added to samples and standards, but may be hazardous to use. It also requires addition of a reagent of an easily ionized compound, such as potassium, to be added to suppress ionization. It is suggested that an air-acetylene flame is more appropriate for routine use. Releasing agents are chemicals which protect the analyte atoms in the flame from forming compounds with other molecular or ionic species, which will depress the absorption in an erratic manner. Either strontium or lanthanum salts are used for this purpose. It is essential that all standard solutions are made up in the same reagent as the samples. This ensures that they not only have any impurities introduced by the reagent solution, but that they have the same viscosity (which can greatly affect the rate of aspiration by the nebulizer) and exert the same interference effect in the flame. A blank solution should always be included, and a control obtained from a bulk sample is good practice for any analysis, and enables one to detect if a systematic error or instrument malfunction should arise. The sample solutions will often require dilution to suit the sensitivity of the particular instrument, however, the sensitivity may be able to be reduced either electronically or by rotation of the burner, and so avoid this extra step. If the standard curve begins to level out towards the horizontal, the flame is probably becoming saturated with the analyte, and dilution is essential.

Wavelengths for AAS. Calcium is measured at 422.7 nm and magnesium at 285.21 nm.

Reagents.

- Calcium stock solution, 1000 µg Ca^{2+} ml^{-1} – stock solutions of many elements for determination by AAS are available commercially. Details for in-house preparation will also be given. Anhydrous calcium nitrate, $Ca(NO_3)_2$, is dried for 1 h at 105°C, then cooled in a desiccator. Transfer 2.05 g to a 100-ml beaker containing water and stir to dissolve. Immediately add 1 ml HCl (36% m/m) to prevent hydrolysis, add with washings to a 500 ml volumetric flask, make up to the mark with water, and mix by shaking.
- Calcium standards, 50 and 0–5.0 µg Ca^{2+} ml^{-1} – pipette 25 ml stock solution into a 500-ml volumetric flask, make up to the mark with M ammonium ethanoate reagent and mix to give a solution of 50 µg Ca^{2+} ml^{-1}. Pipette 0, 0.5, 1, 2.5, 5 and 10 ml of this solution into 100 ml volumetric flasks and make up to the mark with ammonium ethanoate reagent. Standard values are 0, 0.25, 0.5, 1.0, 2.5 and 5.0 µg Ca^{2+} ml^{-1}.
- Releasing agent – dissolve 2.68 g lanthanum chloride heptahydrate $(LaCl_3.7H_2O)$ in water and make up to 100 ml.
- Magnesium stock solution, 1000 µg Mg^{2+} ml^{-1} – dissolve 1.6581 g magnesium oxide (previously dried at 105°C overnight and cooled in a desiccator) in the minimum of hydrochloric acid (approximately 5 M). Dilute with water to 1 l in a volumetric flask to obtain a solution of 1000 µg Mg^{2+} ml^{-1}.
- Magnesium standards, 10 and 0–1 µg Mg^{2+} ml^{-1} – pipette 5 ml stock solution into a 500-ml volumetric flask and dilute to the mark with M ammonium ethanoate reagent to obtain a stock solution of 10 µg Mg^{2+} ml^{-1}. Pipette 0, 2, 4, 6, 8 and 10 ml of the 10 µg Mg^{2+} ml^{-1} stock solution into 100 volumetric flasks and make up to the mark with M ammonium ethanoate reagent and mix. This will give solutions containing 0, 0.2, 0.4, 0.6, 0.8 and 1.0 µg Mg^{2+} ml^{-1}.

Analysis of solutions. Pipette 20 ml of sample and standard solutions into 50-ml beakers, then pipette 1 ml releasing agent solution into each beaker and mix. If readings are off-scale, pipette 5 ml extract plus 15 ml M ammonium ethanoate and the 1 ml releasing agent and retest. Whatever dilution is necessary, ensure the sample plus M ammonium ethanoate solution add up to 20 ml before addition of the 1 ml releasing agent.

Measurement of potassium and sodium by flame photometry

These elements are best determined using flame photometry, as their high atomic emission energy in the flame exceeds their absorption of energy, which results in a higher sensitivity than with atomic absorption spectrophotometry. Check that the appropriate filter is in place for the element being determined, ignite the air–propane flame and ensure an adequate warm-up time. Aspirate the blank solution and adjust the reading to zero. Aspirate the highest standard to allow sensitivity adjustment to give an emission of about 90% maximum reading, and then re-check the zero with the blank. Ensure the standard curve is reasonably linear, then proceed to analyse the samples. Repeat the standards at about 10-min intervals to permit correction for any changes in sensitivity. A quality control sample may be analysed at intervals of about 48 samples.

Reagents.

- Potassium stock solution, 1000 µg K^+ ml^{-1} – weigh 1.293 g potassium nitrate (previously dried for 1 h at 105°C and cooled in a desiccator) into a 100-ml beaker. Dissolve in water, add 1 ml hydrochloric acid (approximately 36% m/m HCl) and 1 drop of toluene, then transfer with washings to a 500-ml volumetric flask, make up to the mark and mix well by shaking.
- Potassium standard solutions, 100 and 0–10 µg K^+ ml^{-1} – pipette 10 ml of the stock solution into a 100-ml volumetric flask and dilute with M ammonium ethanoate reagent to the mark and mix to give a solution of 100 µg K^+ ml^{-1}. Pipette 0, 2, 4, 6, 8 and 10 ml of this solution into 100-ml volumetric flasks and dilute to the mark with M ammonium ethanoate reagent and mix. These will contain 0, 2, 4, 6, 8 and 10 µg K^+ ml^{-1}.
- Sodium stock solution, 1000 µg Na^+ ml^{-1} – weigh 2.542 g sodium chloride (previously dried for 1 h at 105°C and cooled in a desiccator) into a 100-ml beaker. Dissolve in water, add 1 ml hydrochloric acid (approximately 36% m/m HCl) and 1 drop of toluene, then transfer with washings to a 1000-ml volumetric flask, make up to the mark and mix well by shaking.
- Sodium standard solutions, 100 and 0–10 µg Na^+ ml^{-1} – pipette 10 ml of the stock solution into a 100-ml volumetric flask and dilute with M ammonium ethanoate reagent to the mark and mix to give a solution of 100 µg Na^+ ml^{-1}. Pipette 0, 1, 2, 3, 4 and 5 ml of this solution into 100-ml volumetric flasks and dilute to the mark with water and mix. These will contain 0, 1, 2, 3, 4 and 5 µg Na^+ ml^{-1}.

Calculation (1). Results have traditionally been expressed as milliequivalents per 100 g soil. An alternative more recent expression is centimole charge per kilogram soil (cmol$_c$ kg^{-1}), but both expressions give the same numbers. The concentrations of cations using the above methods may be obtained by multiplying the concentration of cation (µg ml^{-1}) in the sample extract solution (obtained by comparing sample readings with the standard curve) by the following factors (plus any dilution factors to bring readings on scale):

Calcium, 0.249; magnesium, 0.412; potassium, 0.128; sodium, 0.2175
Explanation: If a reading of X µg K ml^{-1} is obtained for a solution of 5 g soil in 250 ml extractant (1 in 50 dilution), this amounts to $X/(39.098 \times 10^3)$ milliequivalents K ml^{-1}, or $250X/(39.098 \times 10^3)$ milliequivalents K in 250 ml extractant. This is derived from 5 g soil, thus 100 g soil would contain $(20 \times 250 \times X)/(39.098 \times 10^3) = 0.128$ milliequivalents K.

Thus if 2 g soil were taken instead of 5 g, an additional factor of × 5/2 should be used. If the sample solution for calcium determination was diluted 5 ml solution plus 15 ml M ammonium ethanoate reagent before addition of 1 ml releasing agent, then an additional factor of × 4 will be necessary.

Determination of cation exchange capacity (CEC)

The ammonium extracted by the potassium chloride reagent is analysed by steam distillation. This may be carried out using an automatic instrument such as the Kjeltec Auto 1035 Analyzer (USDA, 1996, pp. 203–210), or a micro (or semi-micro) steam distillation unit such as that described by Bremner and Keeney (1965), or the readily available Markham still. We will describe the manual procedure.

Reagents.

- Ammonium-N standard solution, 140 µg ml^{-1} nitrogen – weigh 0.661 g ammonium sulphate (dried at 105°C for 1 h and cooled in a desiccator) into a 100-ml beaker and dissolve in ammonia-free water (distil deionized water acidified with sulphuric acid), transfer with washings to a 1-l volumetric flask and make up to the mark with the ammonia-free water and mix. This should be stored in a refrigerator, but a quantity allowed to warm to room temperature in a stoppered container before use.
- Boric acid solution, approximately 2% m/v – prepare fresh weekly.
- Mixed indicator – dissolve 0.3 g methyl red and 0.2 g methylene blue in 250 ml ethanol.
- Magnesium hydroxide suspension – heat magnesium oxide (heavy) for 2 h at 800°C. After cooling in a desiccator, make a suspension of 17 g in 100 ml water.
- Octan-2-ol – antifoam agent: use 1 drop when flasks <150–250 ml capacity are used.
- Sulphuric acid, 0.005 M

Procedure. Steam is passed through the steam distillation apparatus for 20–30 min. Check the performance by pipetting 5 ml ammonium-N standard solution into the distillation unit, add 1 drop octan-2-ol, 6 ml magnesium hydroxide suspension and steam distil the released ammonia into 5 ml boric acid solution in a 100-ml conical flask. After approximately 40 ml distillate has been collected over a 5-min period, wash the tip of the condenser into the distillate, add 2–3 drops mixed indicator solution and titrate with 0.005 M H$_2$SO$_4$ until the colour changes from green to purple. A blank distillation/titration is carried out using 5 ml ammonia-free water and subtracted from the standard titre to give a result which should be 5.00 ml.

Pipette a 25-ml (or y ml, where $y \leqslant 50$ ml) aliquot of the soil extract in KCl into the distillation apparatus and proceed as above. If the titre (s) lies outside of the range 0.2–7.0 ml, adjust the volume of extract accordingly. Repeat using a similar aliquot of KCl extractant solution to give a blank titre (b).

Calculation (2).

$$CEC_c = [(s - b) \times 20]/y \text{ cmol}_c \text{ kg}^{-1} \text{ air-dry soil}$$

Determine the percentage moisture content ($z\%$) of the air-dry soil by oven drying overnight at 105°C. Then calculate the above as $CEC_c \times 100/(100 - z)$

cmol$_c$ kg^{-1} oven-dry soil. If the moisture content is relative to the oven-dry soil (z% of oven-dry soil), the calculation becomes:

CEC$_c$ × (100 + z)/100 cmol$_c$ kg^{-1} oven-dry soil.

Notes. Depending on the apparatus, wearing insulating gloves and eye protection, remove the steam lead and then the distillation flask while still hot (or remove the flask, then turn off the steam supply). This is to prevent suckback of flask contents into the steam generator, also to prevent seizure of ground glass joints because of the effect of magnesium oxide.

Other variations include the use of bromocresol green-methyl red indicator, titrating from green through colourless to a pale pink end-point. Magnesium oxide may be used straight from the bottle or replaced by 10 ml 50% m/v NaOH solution. The acid may be 0.01 M HCl (equivalent to 0.005 M H$_2$SO$_4$).

Method 5.3. Determination of effective cation exchange capacity (ECEC)

Reagents.

- Hydrochloric acid, 0.1 M.
- Sodium hydroxide, 0.1 M
- Phenolphthalein indicator, 0.1% (m/v) – dissolve 0.1 g phenolphthalein in 100 ml 95% ethanol.
- Potassium chloride, 1 M – dissolve 74.55 g KCl and make up to 1 l with water.
- Sodium fluoride, 4% (m/v) – dissolve 40 g of NaF in water and make up to 1 l with water.

Procedure. Weigh 20 g of air-dry soil sieved to ≤2 mm into a 250-ml wide-mouth high-density polyethylene screw-cap bottle. The square type bottles fit best the square box of the reciprocating shaker. Add 100 ml 1 M KCl and shake for 15 min. Transfer all the suspension to a filter funnel holding a Whatman No. 6 paper, and collect the filtrate. When leaching has ceased, add two successive 50-ml aliquots of 1 M KCl. Combine the leachates from the total addition of 200 ml (some will be retained in the soil) and mix.

Determination of hydrogen and aluminium

Transfer 100 ml of filtrate to a 250-ml conical flask, add 5 drops of phenolphthalein indicator and titrate to a permanent pale pink colour using 0.1 M NaOH and with alternate swirling and standing. Record the titre that is equivalent to the total acidity (H$^+$ plus Al^{3+}). The associated equations are:

HCl + NaOH = H$_2$O + NaCl

AlCl$_3$ + 3NaOH = Al(OH)$_3$ + 3NaCl

The aluminium hydroxide appears as a hazy white precipitate, and is titrated as described below.

Add one drop of 0.1 M HCl to convert the above pink colour back to colourless, then add 10 ml 4% NaF solution. Titrate with 0.1 M HCl, stirring constantly, until the pink colour just disappears. Next add two more drops of indicator, and if a pink colour returns, titrate again until it disappears and remains colourless for 2 min. This second titration is equivalent to the amount of exchangeable aluminium. The associated equations are:

$$Al(OH)_3 + 6NaF = 3NaOH + Na_3AlF_6$$

$$NaOH + HCl = NaCl + H_2O$$

Determination of exchangeable calcium and magnesium for ECEC

This is carried out by AAS as for CEC above, except that the blank and all standards must be made up using M KCl solution.

Calculation. The ECEC is the sum of the $Ca^{2+} + Mg^{2+} + H^+ + Al^{3+}$ in units of $cmol_c$ kg^{-1} soil (giving the same numerical value as milliequivalents per 100 g soil).

The titration is interpreted as follows:

200 ml M KCl solution ≡ 20 g soil, therefore 100 ml M KCl solution ≡ 10 g soil
1 ml 1.0 M HCl ≡ 1×10^{-3} mol = 0.1 cmol, therefore
1 ml 0.1 M HCl ≡ 0.01 cmol which is per 10 g air-dry soil, therefore
1 ml 0.1 M HCl ≡ 1.0 $cmol_c$ analyte ion per kg air-dry soil

Thus the titre in ml 0.1 M HCl is identical to the concentration of analyte ion(s).

The first titre corresponds to $H^+ + Al^{3+}$, and by subtracting the second titre which corresponds to the exchangeable $cmol_c$ Al^{3+}, one gets the $cmol_c$ H^+.

The Ca^{2+} and Mg^{2+} readings (y) are obtained as μg ml^{-1}. The ratio of soil to extractant is 10:100, therefore 1 kg soil corresponds to 10,000 ml extractant.

y μg Ca^{2+} ml^{-1} ≡ $10_4 \times y$ μg Ca^{2+} kg^{-1} air-dry soil ≡ $10^{-2} \times y$ g Ca^{2+} kg^{-1} air-dry soil

The molar mass of Ca^{2+} is 40.1 g mol^{-1}, thus we get

{$(10^{-2} \times y)/40.1$} $\times y$ mol Ca^{2+} kg^{-1} air-dry soil ≡ $y/40.1$ cmol Ca^{2+} kg^{-1} air-dry soil

Now 1 cmol Ca^{2+} contains 2 $cmol_c$ Ca^{2+}, so the above expression becomes

$2y/40.1$ $cmol_c$ Ca^{2+} kg^{-1} air-dry soil = 0.050 y $cmol_c$ Ca^{2+} kg^{-1} air-dry soil

Correct for soil moisture content as in Method 5.2, Calculation (2), and any other dilution factors.

The expression for exchangeable magnesium is obtained in the same manner, thus

$2y/24.305$ cmol$_c$ Mg^{2+} kg^{-1} air-dry soil = 0.082 y cmol$_c$ Mg^{2+} kg^{-1} air-dry soil

Take into account any correction factors as above.
Sum the values of Ca^{2+} + Mg^{2+} + H$^+$ + Al^{3+} to obtain the ECEC in units of cmol$_c$ kg^{-1} soil.

Method 5.4. Determination of fulvic and humic acids

Microorganisms break down plant and animal residues in the soil to form a stable dark brown organic material called humus. It is composed of a mixture of large complex molecules (molar masses 20,000–100,000 g mol^{-1}). Lignin-type precursors result in benzene ring (aromatic) compounds substituted with hydroxyl (—OH), methoxyl (—OCH$_3$) and carboxyl (—COOH) groups such as gallic and vanillic acids. These react by oxidation and/or polymerization to form dark brown substances (Flaig, 1997) typical of water leached through peat.

The main sources of negative charge on the humus particle arise from the —COOH and —OH (phenolic) groups, with only the —COOH being significantly charged below pH 7. The sources of these charges may be mainly separated into the fulvic and humic acid fractions. These are not distinct chemical species, but rather two groups of complex organic soil substances with some common chemical characteristics, which are separated and differentiated purely by their solubility in sodium hydroxide and then hydrochloric acid under the stated conditions. Aiken *et al.* (1985) have defined fulvic acids as 'the fraction of humic substances that is soluble under all pH conditions', and humic acids as 'the fraction of humic substances that is not soluble in water under acid conditions, but becomes soluble at greater pH'. In general, fulvic acid has a lower molecular weight, lower N content and is possibly more aromatic than humic acid. With spectroscopic techniques, IR shows broad bands from functional groups, but little information on the composition of humic substances (HS); UV-Vis using 465/665 nm ratios shows differences depending on source, but no meaningful chemical interpretation of the spectra; NMR has yielded more information and shown the persistence of lignin- and tannin-type residues. Wet chemical and some spectroscopy procedures have been reviewed by Hayes and Swift (1978), Swift (1996) and Hayes (1997).

The ratio of fulvic to humic acid varies between soils and between horizons of the same soil. Humic fractions are involved with solubilization of the sesquioxides (especially gibbsite, Al(OH)$_3$; goethite, FeOOH; haematite, Fe$_2$O$_3$; and ferrihydrite, Fe$_2$O$_3$.nH$_2$O). It is therefore desirable to determine the Al and Fe associated with these fractions. The scheme of separations is shown in Fig. 5.1.

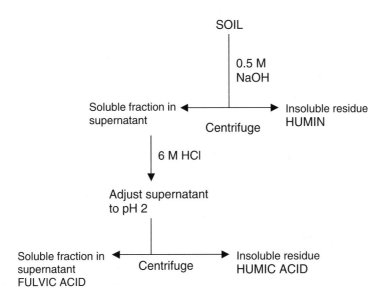

Fig. 5.1. Scheme of separating the soil fulvic and humic acid fractions.

Procedure. Weigh 10 g air-dry sieved (≤2 mm) earth into a 250-ml plastic centrifuge bottle, add 200 ml 0.5 M NaOH and shake overnight. Centrifuge at 2000 rpm for 20 min to allow sedimentation of the insoluble humin and decant all the supernatant into a clean centrifuge bottle. Adjust to pH 2.0 with 6 M HCl, then centrifuge at 2000 rpm for 20 min to cause sedimentation of the humic acid. Decant the solution of fulvic acid into a 250-ml volumetric flask. Wash the sediment of humic acid with 30 ml 0.5 M HCl, centrifuge, add supernatant to the volumetric flask, and make up to the mark with water and mix. Read the optical density at 465 nm, diluting if necessary to bring on scale. The approximate concentration of fulvic acid in mg l^{-1} is given by comparing with the graph of optical density vs. concentration (Professor W.A. Adams, Aberystwyth, 2001, personal communication) as shown in Fig. 5.2. (The fulvic acid is in a solution of NaCl, therefore gravimetric determination is not possible.) It is recommended, however, that the extinction (absorption) coefficient should be determined for humic substances from different origins, from the modified Beer–Lambert expression (Schnitzer and Khan, 1972):

$$E^{0.001\%}_{1\,cm} = OD/cl$$

where E is the extinction coefficient, OD is the optical density, c is a 0.001% solution of the humic compound in the stated reagent, and l is the internal cell length of 1 cm.

Wash the humic acid residue with 200 ml 0.5 M HCl, centrifuge and discard the supernatant. Wash the humic acid out of the centrifuge bottle with 60% IMS into a pre-weighed oven-dry 100 ml glass beaker. Evaporate to dryness carefully on a hotplate, avoiding loss by spitting, cool in a desiccator

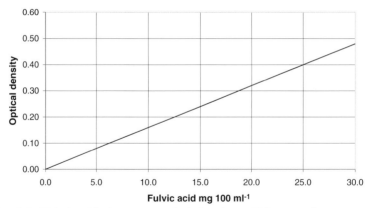

Fig. 5.2. Relationship between optical density at 465 nm and concentration of ash-free sedge peat fulvic acid.

and reweigh. The difference in weights gives the weight of humic acid plus ash.

Ignite in a muffle furnace at 500°C overnight to burn off the humic acid fraction, cool in a desiccator (leave lid slightly open for the first minute to allow air to expand) then reweigh the beaker containing the residual ash. Subtract this weight from the weight of beaker and residue before ashing to obtain the weight of ash-free humic acid.

Add 20 ml 6 M HCl dropwise on to the ash and warm carefully to dissolve, and then make up with water to 50 ml in a volumetric flask. Using suitable dilutions where necessary, analyse for Ca, Fe and Al by AAS. It will be necessary to use a nitrous oxide-acetylene flame for aluminium, otherwise use the titration method as in Method 5.3.

Calculation. Ash-free fulvic acid y mg 100 ml^{-1} is read from the chart. This solution resulted from 10 g air-dry soil in 250 ml solution. Therefore 250 ml solution contains $y \times 250/100$ mg fulvic acid from 10 g air-dry soil, which converts to $25y$ mg fulvic acid 100 g^{-1}, or $0.025y\%$ air-dry soil. This must be multiplied by any dilution factor before reading the optical density, also converted to percent oven-dry soil (see Method 5.2, Calculation (2)).

The weight of ash-free humic acid was derived from 10 g air-dry soil, therefore should be multiplied by 10 to convert to percent air-dry soil and further converted to percentage oven-dry soil. The fulvic and humic acid content may be expressed as a percentage of total soil organic matter, which is quantified as soil organic carbon (SOC). SOC is conveniently determined from the loss on ignition, where the correlation is:

Loss on ignition = 1.94 SOC + 16.0 (r^2 = 0.96) McGrath (1997)

The ratio of fulvic to humic acid is also significant; both, together with carbohydrates (uronic acids and sugars), are linearly related to the SOC. The humic acid increases and the fulvic acid decreases as the SOC increases (McGrath, 1997).

The Al, Ca and Fe may be expressed as a percentage of the humic acid content.

Discussion 5.5. Determination of available nitrogen

Soil nitrogen reactions

Fixation:	$N_2 \rightarrow NH_4^+ \rightarrow$ organic-NH_2
Immobilization:	NH_4^+ and $NO_3^- \rightarrow$ organic-NH_2
Mineralization:	organic-$NH_2 \rightarrow NH_4^+$
Nitrification:	$NH_4^+ \rightarrow NO_2^- \rightarrow NO_3^-$
Volatilization:	$NH_4^+ + OH^- \rightleftharpoons NH_3\uparrow + H_2O$

The available nitrogen is equivalent to the mineralized nitrogen, which consists of the soluble nitrate and nitrite, and the exchangeable and soluble ammonium nitrogen. These compounds fluctuate over short periods and are greatly affected by microbial activity; the ammonia gas may escape from the sample by volatilization. The soil sample should be transported in an icebox and transferred to a freezer unless immediately analysed. If just the available nitrate is required, it may be analysed according to the colorimetric auto-analysis method given for nitrate in herbage (Method 7.6). It may also be extracted in the traditional manner using either a saturated or a 0.01 M calcium sulphate solution to reduce cloudiness in the filtrate, followed by analysis using a colorimetric method or a selective ion nitrate electrode. The latter method is given below.

Method 5.5a. Determination of nitrate by selective ion electrode

Reagents.

- Calcium sulphate solution, 0.01 M – dissolve 1.72 g $CaSO_4.2H_2O$ in water and make up to 1 l.
- Nitrate stock standard solution, 0.2 mg ml^{-1} nitrate-N – dissolve 0.722 g of potassium nitrate in 0.01 M $CaSO_4$ solution, dilute to 500 ml and store in a refrigerator.
- Nitrate working standard solution, 1–100 µg ml^{-1} nitrate-N – pipette 0.5, 2, 5, 10, 25 and 50 ml of the stock standard into 100-ml volumetric flasks and make up to the mark with 0.01 M $CaSO_4$ solution and mix.

Procedure. Transfer a 20-ml level scoop of air-dry soil sieved to $\leqslant$2 mm to a 125-ml wide-mouth HDPE screw-cap bottle, add 50 ml 0.01 M calcium sulphate solution, and shake for 15 min on a reciprocating shaker (MAFF/ADAS use 275 strokes per minute, Northeastern US use 200 oscilla-tions per minute). Filter through a 125-mm Whatman No. 2 filter paper into a sample bottle, discarding the first few millilitres, which may contain a trace of nitrate from the paper. Filter a solution of extractant in the same manner to provide a blank.

Set up the meter and first measure the lowest standard, allowing sufficient time for a steady reading to be obtained. Record the indicated millivolts, and proceed to analyse the remaining standards and sample solutions. Ensure standards and sample solutions are at the same temperature, which should be noted for reference. The standard graph is prepared on semi-logarithmic paper by comparing mV readings on the linear axis to µg ml^{-1} nitrate-N on the logarithmic axis. Values of the order of 140 mV for 1 µg ml^{-1}, and 30 for 100 µg ml^{-1} nitrate-N may be expected.

Calculation. There were 20 ml soil in 50 ml of extractant, therefore sample concentration values from the standard graph should be multiplied by 2.5, the results being expressed as mg l^{-1} nitrate-N calcium sulphate extractable in air-dry soil. Include any extra dilution factors, and, if required, convert to oven-dry soil using the appropriate factor, as in Method 5.2, Calculation (2).

A separate bulk-density determination enables results to be expressed in terms of mg kg^{-1} soil.

Discussion 5.5b. Determination of total mineralized nitrogen

Some workers use 20 g air-dry soil, add 12 ml water and incubate in the dark at 22°C for 28 days, making good any moisture loss. Others extract the air-dry soil directly. The given method uses the fresh (or thawed) soil sample. The usual extractant for moist soils is 2 M KCl, but for dry soil, 10% m/v KCl (1.34 M) may be used. The amount of nitrite-N in the soil is usually so small that it can be neglected, but it will be incorporated in the amount of nitrate recorded.

There are two approaches to the analysis. In the first, the extractable ammonium-N is determined by a Kjeldahl distillation, and this is subtracted from the value for ammonium plus nitrate (and nitrite) determined by a further separate Kjeldahl distillation preceded by reduction with nascent hydrogen produced by the reaction of Devarda's alloy (45% Al, 50% Cu, 5% Zn) in strongly alkaline solution:

production of hydrogen: $2Al + 2OH^- + 2H_2O = 2AlO_2^- + 3H_2\uparrow$

overall reaction: $3NO_3^- + 8Al + 5OH^- + 2H_2O = 8AlO_2^- + 3NH_3$

The second, quicker, way is simply to add the Devarda's alloy to the distillation flask after the first distillation, and then redistil the ammonia produced from the reduced nitrate – this is described below.

Method 5.5b.i. Determination of extractable ammonium-N

Reagents.

- Ammonium-N standard solution, 140 μg ml⁻¹ nitrogen – weigh 0.661 g ammonium sulphate (dried at 105°C for 1 h and cooled in a desiccator) into a 100-ml beaker and dissolve in ammonia-free water (distil deionized water acidified with sulphuric acid), transfer with washings to a 1-l volumetric flask and make up to the mark with the ammonia-free water and mix. This should be stored in a refrigerator, but a quantity allowed to warm to room temperature in a stoppered container before use.
- Boric acid solution, approximately 2% m/v – prepare fresh weekly.
- Mixed indicator – dissolve 0.3 g methyl red and 0.2 g methylene blue in 250 ml ethanol.
- Magnesium hydroxide suspension – heat magnesium oxide (heavy) for 2 h at 800°C. After cooling in a desiccator, make a suspension of 17 g in 100 ml water.
- Octan-2-ol – antifoam agent: use 1 drop when flasks < 150–250 ml capacity are used.
- Potassium chloride, 2 M – dissolve 745.51 g KCl in water, and make up to 5 l.
- Sulphuric acid, 0.005 M

Procedure. Steam is passed through the Markham steam distillation apparatus for 20–30 min. Check the performance by pipetting 5 ml ammonium-N standard solution into the 25 ml reservoir of the distillation unit, add 1 drop octan-2-ol, 6 ml magnesium hydroxide suspension and steam distil the released ammonia into 5 ml boric acid solution in a 100-ml conical flask. After approximately 40 ml distillate has been collected over a 5-min period, wash the tip of the condenser into the distillate, add 2–3 drops mixed indicator solution and titrate with 0.005 M H_2SO_4 until the colour changes from green to purple. A blank distillation/titration is carried out using 5 ml ammonia-free water and subtracted from the standard titre to give a result which should be 5.00 ml.

Pipette a 25 ml (or y ml, where $y \leqslant 50$ ml) aliquot of the soil extract in KCl into the distillation apparatus and proceed as above. If the titre (s) lies outside of the range 0.2–7.0 ml, adjust the volume of extract accordingly. Repeat using a similar aliquot of KCl extractant solution to give a blank titre (b).

Calculation. Subtract the blank from the soil titre, multiply the difference by 700 and divide by y, to get the mg kg⁻¹ ammonium-N, potassium chloride extractable, in air-dry soil. Express results for oven-dry soil as in Method 5.2, Calculation (2).

Method 5.5b.ii. Determination of extractable nitrate-N

Reagents.

- Boric acid solution, approximately 2% m/v – prepare fresh weekly.
- Devarda's alloy – finely ground powder ($\leq$ 53 μm).
- Magnesium hydroxide suspension – heat magnesium oxide (heavy) for 2 h at 800°C. After cooling in a desiccator, make a suspension of 17 g in 100 ml water. Alternatively use MgO powder.
- Mixed indicator – dissolve 0.3 g methyl red and 0.2 g methylene blue in 250 ml ethanol.
- Nitrate-N standard solution, 0.14 mg ml^{-1} nitrate-N – dissolve 1.011 g potassium nitrate (oven-dry) in water, transfer with washings to a 1–l volumetric flask, make up to the mark and mix.
- Octan-2-ol – antifoam agent: use 1 drop when flasks < 150–250 ml capacity are used.
- Sulphuric acid, 0.005 M

Procedure. Pipette 5 ml nitrate-N standard solution into the distillation flask, add 1 drop octan-2-ol, approximately 0.5 g Devarda's alloy, 6 ml magnesium hydroxide suspension (or 0.5 g MgO), and steam-distil the ammonia into 5-ml boric acid solution in a 100-ml conical flask. After approximately 40 ml distillate has been collected over a 5-min period, wash the tip of the condenser into the distillate, add 2–3 drops mixed indicator solution and titrate with 0.005 M H$_2$SO$_4$ until the colour changes from green to purple. Carry out a blank distillation using 5 ml water instead of extract solution and subtract from the standard titration to give a difference of 5.0 ml.

After determination of extractable ammonium-N (5.4b.i), add approximately 0.5 g Devarda's alloy and distil as above. Adjust the size of the aliquot of extract if the titre lies outside the range 0.2–7.0 ml.

Calculation. Subtract the blank titre from that of the soil extract, multiply by 700 and divide by *y* (see Method 5.5b.i) to give the mg kg^{-1} nitrate-N, potassium chloride extractable, in the air-dry soil. Express results for oven-dry soil as in Method 5.2, Calculation (2). Sum the extractable ammonium-N and nitrate-N to obtain the total mineralizable nitrogen in the soil.

Discussion 5.6. Determination of organic plus ammonium nitrogen

Refer to Chapter 4 for a discussion on 'Total soil nitrogen'. Instructions for using the Foss Tecator Kjeltec Auto 1030 Analyzer are given in USDA, 1996, pp. 535–538.

Method 5.6a. Determination of soil nitrogen by autoanalysis

If the analysis is to be carried out in a similar way to total N in herbage, by acid-digestion in test tubes in an aluminium block followed by a colorimetric autoanalysis procedure, then take a 0.5-g sample of air-dry soil. If serious frothing occurs, take 0.2 g soil, and adjust the calculation accordingly. Reducing the amount to ⩽0.1 g may give peaks which are too small to be read with confidence. See Chapter 7 (pp. 138–141) for details of standards, reagents and method.

If the soil contains significant amounts of nitrate, which should be included with the organic and ammonium-N for total nitrogen, it must be reduced to ammonium-N by the following procedure (Method 5.6a.i). Determination of N on duplicates of the same soil sample with and without a reducing step will enable an estimation of the nitrate-N (plus any nitrite-N) by difference. A more precise colorimetric determination of just the nitrate component uses the method described for herbage, except that 10 ml of fresh or freshly thawed soil is used.

Method 5.6a.i. Reduction of nitrate before digestion and colorimetric autoanalysis

Reactions involved

1. Sulphuric acid converts *nitrate* to *nitric acid*
2. Salicylic acid converts *nitric acid* to *nitrosalicylic acid*
3. Zinc dust converts *nitrosalicylic acid* to *aminosalicylic acid*
4. Hot sulphuric acid converts *aminosalicylic acid* to *ammonium sulphate*, which is the same form to which protein nitrogen is converted.

Reagents.

- Salicylic acid, $2\text{-HOC}_6\text{H}_4\text{COOH}$.
- Sulphuric acid–selenium reagent, approx 98% m/m H_2SO_4 – *Safety note* wear PPE because this is highly corrosive. Add 4 g selenium powder to 400 ml sulphuric acid (approx. 98% m/m H_2SO_4), and heat until just fuming. Stir occasionally with a glass rod until all the selenium has dissolved to give a dark green solution. Carefully remove from the hotplate and allow to cool. Make up to approximately 1 l with sulphuric acid.
- Zinc dust.

Procedure. Weigh 0.5 g air-dry soil (0.1000 g for herbage samples) into the digestion tube. Add 5 ml sulphuric acid containing 0.4% w/w selenium catalyst, followed by 0.10 g salicylic acid and mix by gentle agitation of the tube. After allowing to stand for 30 min, add approximately 0.1 g zinc dust. Mix by gentle agitation, and allow sufficient time for the hydrogen gas to evolve. This is best carried out in a fume cupboard, as the fumes are unpleasant. Follow by heating at 100°C until frothing ceases, and then heat at 310°C for 4.25 h. Analyse according to the procedure for herbage nitrogen in Method 7.7a.

Notes:

1. After diluting the Kjeldahl digest to give a 50% aqueous solution, one observes a residue of silica and a fine white mineral deposit at the bottom of the digestion tube. There may also be some cloudiness in the solution, which should be left to settle for 48–72 hours. Using a disposable pasteur pipette, transfer sufficient of the upper clear solution to almost fill the autoanalyser sample cup.
2. During analysis, a flocculent light brown precipitate may be produced, which can build up in the flowcell, causing noisy peaks and drifting. A dialyser module could prevent this situation; otherwise the flowcell could be 'scrubbed' by forcing an air-bubble through between samples.
3. The Kjeldahl digestion can convert up to 20% of any nitrate-N to ammonium-N. This would usually only increase the ammonium-N by an insignificant amount and may therefore be ignored.
4. The volume of residue in the graduated digestion tube has been measured for a number of soils. It appears bulky, but after filtration, washing with 90% ethanol, drying and transferring to a 5-ml measuring cylinder, and pipetting in 2 ml ethanol, the volume was found to vary between 0.20 and 0.30 ml for all types of temperate and organic soils. The residue means the soil organic-N is contained in 9.7–9.8 ml rather than 10 ml, and therefore the readings are reduced by dividing the mg l^{-1} by 510–515 (or multiplying by 0.00196–0.00194) where the units are % N. We have adopted the average value of 513 or 0.00195 respectively. Where the units are required as mg N g^{-1} or g N kg^{-1}, these factors become 51.3 and 0.0195 respectively.

Calculation. For 0.5 g air-dry soil sample, digested and made up to 10 ml, the concentration in that solution is read from the standard graph as *y* mg N l^{-1}. The concentration of organic-N in the air-dry soil is expressed by:

$y \times 0.00195$ (or $y/513$) % N in air-dry soil.

Express results for oven-dry soil as in Method 5.2, Calculation (2).

Method 5.6b. Determination of organic plus ammonium-N by digestion and distillation[a]

Nitrogen in the sample is converted to ammonium-N by digestion with sulphuric acid and sodium sulphate with a copper–selenium catalyst. The ammonia liberated with sodium hydroxide is removed by steam distillation and determined titrimetrically.

[a](MAFF/ADAS, 1986, pp. 150–151 with Crown Copyright Permission).

Digestion stage

Apparatus.

- Kjeldahl flasks – 300 ml.
- Macro-Kjeldahl digestion unit – with adjustable heating.

Reagents.

- Copper–selenium catalyst tablets – each tablet contains 0.5 g of cupric sulphate pentahydrate, and 0.02 g of selenium.
- Sodium sulphate tablets – each tablet contains 2.5 g of anhydrous sodium sulphate.
- Sulphuric acid, approximately 98% m/m H_2SO_4.

Procedure. Transfer 5 g of air-dried soil, ground to pass a 2-mm mesh sieve, into a Kjeldahl flask and add four sodium sulphate tablets and one copper–selenium tablet. Add 25 ml sulphuric acid. Swirl the acid until no particles of the sample adhere to the bottom of the flask.

Heat the flask gently until frothing ceases, more strongly until the solution clears, and then for a further 1 h with the sulphuric acid condensing in the lower part of the neck of the flask. Allow to cool, carefully add approximately 100 ml of water and warm to dissolve the soluble material. When cool, transfer quantitatively into a 250-ml volumetric flask and dilute to 250 ml. Retain the diluted digest for the determination of ammonium nitrogen.

Carry out a blank determination using 2 g of sucrose in place of the sample.

Distillation stage

Apparatus.

- Distillation unit – the Markham semi-micro distillation unit is suitable.

Reagents.

- Ammonium-N standard solution, 0.28 mg ml^{-1} of nitrogen – dry ammonium sulphate at 102°C for 1 h and cool in a desiccator. Dissolve 1.321 g of the dried salt in water and dilute to 1 l.
- Boric acid solution, approximately 1% m/v.
- Methyl red–methylene blue solution – dissolve 1.25 g of methyl red and 0.825 g of methylene blue in 1 l of ethanol.
- Sodium hydroxide solution, 50% m/v.
- Sulphuric acid 0.01 M.

Procedure. Steam out the distillation unit for 20 min. Pipette 5 ml of ammonium-N standard solution into the unit. Add 7 ml of sodium hydroxide solution and steam distil the liberated ammonia into 5 ml of boric acid solution. Collect 20 ml of distillate. Add 2–3 drops of methyl red–methylene blue solution and titrate with 0.01 M sulphuric acid until the green colour changes to purple. Carry out a blank determination using 5 ml of water in place of the ammonium-N standard. The titre of the ammonium-N standard minus the distillation blank should be 5.00 ml.

Pipette 10 ml of the sample digest (see under 'Digestion stage') into the distillation unit, and proceed with addition of sodium hydroxide solution, distillation and titration as above.

Calculation. Subtract the titre given by the blank from that given by the sample digest. Multiply the difference by 1.4. The result gives the g kg^{-1} of nitrogen in the sample. (Multiply the difference by 0.14 to get the % N in the sample.) If required, express results for oven-dry soil as in Method 5.2, Calculation (2).

Discussion 5.7. Determination of soil organic matter

An approximate determination of soil organic matter is achieved by burning it off in a muffle furnace and measuring the weight loss. This 'loss on ignition' may overestimate the true content, especially with clay soils. The —OH groups of clay minerals dehydroxylate, as also iron and aluminium oxyhydroxides and aluminium hydroxide, yielding H_2O and therefore losing weight. Illite may lose structural water at temperatures around 150–250°C. Calcium carbonate loses CO_2 when heated at 550°C to form CaO (Miall and Miall, 1956), therefore as low a temperature as possible is chosen (400°C), which will still achieve the oxidation of the organic matter.

Method 5.7a. Determination of soil organic matter by loss on ignition

Procedure. Weigh approximately 10 g (±0.001 g) air-dry soil, sieved to ≤2 mm, into an accurately weighed, oven-dried, 100-ml beaker. Dry at 105°C overnight, cool in a desiccator, and weigh beaker plus sample. Place the beaker in a cold muffle furnace, switch on and ignite at 400°C overnight. Transfer to an oven at 105°C for 15 min, then to a desiccator until cool, before weighing the beaker plus ignited soil.

Calculation. The percentage loss on ignition is calculated as:

$$\frac{\text{oven-dry soil weight} - \text{ignited soil weight}}{\text{oven-dry soil weight}} \times 100\%$$

The weight of water (W) in the air-dry soil may also be calculated as:

(wt of beaker + air-dry soil) – (wt of beaker + oven-dry soil)

The weight of air-dry soil (S) is:

(wt of beaker + air-dry soil) – (wt of beaker)

The percentage of water in the air-dried soil is therefore:

$$\frac{W}{S} \times 100\%$$

and the percentage of water relative to the oven-dry soil becomes:

$$\frac{W}{(\text{wt of beaker} + \text{oven-dry soil}) - (\text{wt of beaker})} \times 100\%$$

Method 5.7b. Determination of easily oxidizable organic C by Tinsley's wet combustion

If the soil organic matter contains an element in a fairly constant percentage, then the determination of that element should enable deduction of the organic matter. This has been attempted by determining nitrogen, but this was found to vary too widely in organic soil materials. More commonly, carbon has been determined. It is assumed that soil organic matter contains on average 58% carbon, so that multiplication of the value for carbon by 1.72 (100/58) will give the percentage organic matter. The other factor is the percentage organic matter to be oxidized by the chromic acid produced *in situ*, with or without external heat. The average for the classic Walkley and Black (1934) method is 77%, which may not be valid for subsoils (Allison, 1960), whereas Tinsley's wet combustion method (Tinsley, 1950) oxidizes most of the organic matter, facilitated by the presence of perchloric acid. Carbon in graphite and coal may not be oxidized. It is therefore preferable to report the 'percentage easily oxidizable organic C' rather than the 'percentage organic matter'.

Errors. Chromic acid digestion can incur errors from any significant presence of chlorides, which can reduce the dichromate ion to give high results. The addition of silver sulphate to the digestion acid, or prior leaching of the soil could eliminate this effect. Any ferrous ion (from an anaerobic soil situation) would also reduce dichromate, but the aerobic soil drying process would normally oxidize this to ferric ion. Higher oxides of manganese could compete with dichromate in oxidizing organic matter, but this is not usually significant. Some dichromate will be reduced by organic hydrogen, but this is approximately compensated for by the loss of carbon as CO_2 from organic oxygen:

$$RCOOH \rightarrow RH + CO_2$$

Safety. The addition of concentrated sulphuric acid on sodium (or potassium) dichromate is to produce chromium trioxide, which is a powerful oxidizing agent capable of oxidizing carbon to carbon dioxide. The solution is loosely called chromic acid, but although true chromic acid H_2CrO_4 has not been isolated, the aqueous solution contains dichromic acid, $H_2Cr_2O_7$. The acid–dichromate reagent (hexavalent chromium) is corrosive to skin, respiratory and gastrointestinal tract, and may create a cancer risk. There may be restrictions on disposal into municipal sewerage systems, therefore storage for professional chemical waste disposal may be required. Adequate PPE should be employed to protect eyes, skin and lungs. With adequate supervision, the procedure has been used in practical classes for many years with no reportable incident.

Reaction equations. Oxidation step: dichromate ($Cr_2O_7^{2-}$) and organic carbon (C^o):

$$2Cr_2O_7^{2-} + 3C^o + 16H \rightarrow 4Cr^{3+} + 3CO_2 + 8H_2O$$

Titration step: ferrous ion (Fe^{2+}) with excess dichromate:

$$6Fe^{2+} + Cr_2O_7^{2-} + 14H^+ \rightarrow 2Cr^{3+} + 6Fe^{3+} + 7H_2O$$

Reagents.

- Acid-dichromate mixture, 66.7 mM – dissolve 20.0 g sodium dichromate dihydrate ($Na_2Cr_2O_7.2H_2O$, mol. wt 298.00, equivalent wt 49.67) in about 50 ml water in a 2-l beaker and carefully add 400 ml sulphuric acid (98% m/m) with stirring, and allow to cool. Next add 140 ml perchloric acid 70% (1.70 g ml^{-1}) or 163 ml of 60%. Make up to 1 l with water.
- Ferrous ammonium sulphate, approximately 0.4 M – carefully add 5 ml sulphuric acid (98% m/m H_2SO_4) to 1.5 l water, stir slowly to dissolve, then add 314 g ammonium ferrous sulphate [(NH_4)$_2SO_4.FeSO_4.6H_2O$, mol. wt 392.14] stir to dissolve and dilute to 2 l.
- Ferroin indicator – slowly dissolve 3.71 g of o-phenanthroline and 1.74 g $FeSO_4.7H_2O$ in 250 ml water.

Procedure. Grind a sample of soil, previously sieved to ≤2 mm, in a pestle and mortar, and sieve to ≤0.5 mm, or ≤30 mesh (US methods use the 60 mesh Market Grade Testing Sieve, 0.23 mm mesh opening). Sufficient soil to contain 10–20 mg carbon should be accurately weighed into a 500 ml conical (Erlenmeyer) flask with ground glass neck to fit a cold finger condenser. (Typical amounts of sample are: surface soil, 1 g; subsoil, 4 g; organic soil, 0.1 g; compost, 0.1 g and 40 ml dichromate mixture.)

Pipette, using a suitable safety pipette filler, 25 ml acid–dichromate mixture into the flask, fit the cold finger dispenser and slowly turn on the water supply. Place on a hotplate and simmer for 2 h.

While the sample is refluxing, standardize the ferrous ammonium sulphate against the dichromate mixture. This must be done daily. Carry out two successive blank digestions with heating for 10 min using 25 ml acid–dichromate, but no soil. Cool and add approximately 100 ml water followed by 4 drops of Ferroin indicator. Titrate with the 0.4 M ferrous ammonium sulphate until the indicator changes from blue-green to reddish-grey. If this is overshot, pipette 1 ml acid–dichromate and titrate dropwise to the end-point, allowing for the extra dichromate in the calculation step.

After digesting the soil samples, allow to cool and titrate in the same way. If more than 20 ml of acid–dichromate have been reduced, repeat using less sample.

Calculation

1. Calculate the number of mol of $K_2Cr_2O_7$ in 25 ml acid–dichromate reagent using the equation: no. mol × concn. (mol l^{-1}) × vol. (l)
= 0.0667 × 25/1000 = 1.668 × 10^{-3} mol, or 1.668 mmol.
2. If the standardization titration was 26 ml ferrous ammonium sulphate solution, it had reacted with 25 ml acid–dichromate containing 1.668 mmol. If the back-titration of the residual acid–dichromate was 10 ml ferrous ammonium sulphate solution, then the residual acid–dichromate amounted to 10 × 1.668/26 = 0.642 mmol. The amount used to oxidize the organic carbon is therefore
1.668–0.642 = 1.026 mmol.
3. From the reaction equation for the oxidation step, it can be seen that 3 mol of carbon react with 2 of dichromate. The number of mol of C oxidized is therefore 1.026 × 3/2 = 1.539 mmol. Since the molar mass of C is 12.0, the mass of C oxidized is therefore 1.539 × 10^{-3} mol × 12 g mol^{-1}, or 18.5 mg. If this is in 1 g soil, the easily oxidizable organic C content becomes 18.5 mg g^{-1} air-dry soil. This can be expressed for oven-dry soil as in Method 5.2, Calculation (2).
4. The calculation above reduces to the following equation:
C content (mg C g^{-1} air-dry soil) = 30(1–x/y)/soil mass,
where x = residual acid–dichromate titre, and y = standardization titre.
5. The organic matter may be approximately obtained by multiplying by 1.74 as explained above. In the calculation example, this would become:
18.5 × 1.72 = 31.82 mg organic matter g^{-1}, or 3.18%.

Discussion 5.8. Determination of pH and lime requirement

Definition. Soil pH is defined as the negative logarithm (to the base of 10) of the H^+ activity in the soil solution measured under the stated conditions.
    The activity approximates to the concentration [H^+] in mol l^{-1} for dilute solutions, thus:

$$pH = -\log_{10} [H^+] \text{ or } \log_{10} 1/[H^+]$$

Acids dissociate to yield hydrogen ions, which will reduce the pH value because 1/[H^+] becomes smaller. Alkalis supply hydroxyl ions (OH^-), which associate with hydrogen ions, thus [H^+] decreases and 1/[H^+] increases. Pure water partially dissociates into hydrogen and hydroxyl ions: $H_2O \rightleftharpoons H^+ + OH^-$ where [H^+][OH^-] = 10^{-14} mol l^{-1}, and is termed the dissociation constant, K_w. Thus [H^+] = [OH^-] = 10^{-7} = pH 7 for a neutral solution. Soils with pH values < 7 are referred to as *acid*, and > 7 as alkaline, and if the pH is 7, as *neutral*.
    Pure water in equilibrium with atmospheric CO_2 has a pH of 5.6. If a soil pH is lower or higher than this, it is acting as an acid or base respectively. Several soil components act as buffers (hydroxy aluminium monomers or polymers, soil organic matter and undissolved carbonates), therefore lime requirement tests may also be required.

Method 5.8a. Measurement of pH

Reagents.
• Buffer capsules/tablets – dissolve to make solutions of pH 4.0, 7.0 and 9.0.

Procedure. Calibrate the pH meter according to the manufacturer's instructions using buffers to cover the pH range of the soil samples. Transfer a 10-ml scoop of sieved (⩽2 mm) air-dry soil (struck off level without tapping) into a flat-bottom plastic vending cup, add 25 ml water and a magnetic PTFE-encased stirrer bar. Place on a multi-position electronic stirrer unit (e.g. 15-place Variomag) and stir for 15 min. Tilt the cup, if necessary, to ensure the pH electrode is sufficiently immersed (the soil suspension should reach the porous plug liquid junction on the side of a combination glass electrode). The electrode should not be abraded by the abrasive soil at the bottom of the cup. Swirl a couple of times and allow the drift in pH to stabilize (about 30 s) before taking the reading. Rinsing between samples is not necessary unless soils have widely differing pH values. Recalibrate the meter hourly. If required, retain the suspension for the determination of lime requirement.

Method 5.8b. Determination of lime requirement

Definition. The lime requirement of a mineral soil is the number of tonnes of calcium carbonate calculated to raise the pH of a hectare of soil 200 mm deep, under field conditions, to, and maintain at, 6.5.

A low pH indicates that lime is required, but not the quantity. Excess is not only wasteful, but may render certain elements (e.g. Fe, Mn, and B) unavailable to plants. There are several methods for determining the lime requirement, including adding excess calcium bicarbonate and back-titrating the excess; adding increasing amounts of calcium hydroxide and monitoring the pH; and the use of a buffer solution (MAFF/ADAS, 1986, pp. 150–151), which will be described below.

Reagents.
• Buffer solution, double strength – add to 4.5 l water, 400 g oven-dried calcium acetate, 80 g 4-nitrophenol and 6 g of light magnesium oxide. [*Safety note*: 4- (or para-) nitrophenol may cause eye irritation or irreversible eye injury, and may be harmful by absorption through skin or ingestion.] Heat the mixture to dissolve the solids and dilute to 5 l. Filter if the solution is not clear. The pH of this solution should lie between 6.9 and 7.1; adjust by the addition of hydrochloric acid or magnesium oxide as necessary.
• Buffer solution, single strength – add 1 volume of double strength buffer solution to 1 volume of water and mix. The pH of this solution should lie between 6.9 and 7.1; if necessary, adjust as above.

Procedure for mineral soils of pH 5.0 to 6.4 inclusive. Add 20 ml of single strength buffer solution to the soil suspension retained from the pH determination, and stir for 5 min. Mix 25 ml water with 20 ml single strength buffer with pH adjusted as above and recalibrate the pH meter to indicate pH 7.00 using this solution. Measure the pH of the stirred sample. If the pH is < 6.0, again stir a fresh sample of 10 ml soil with 25 ml water for 15 min and proceed as for mineral soils of pH < 5.0 as described below.

Calculation for mineral soils of pH 5.0–6.4. Subtract the indicated pH from 7.00 and multiply by 11.2. The result gives the lime requirement as tonnes ha^{-1} calcium carbonate.

Procedure for mineral soils of pH less than 5.0. Add 20 ml of double strength buffer solution to the soil suspension retained from the pH determination, and stir for 5 min. Mix 25 ml water and 20 ml double strength buffer with pH adjusted to between 6.9 and 7.1, and use to recalibrate the pH meter to read 7.00. Read the pH of the stirred sample.

Calculation for mineral soils of pH less than 5.0. Subtract the indicated pH from 7.00 and multiply by 22.4 to obtain the lime requirement expressed as tonnes ha^{-1} calcium carbonate.

Notes on the calculation. A full explanation of soil buffer capacity and derivation of the above factors is given by Rowell (1994, pp. 171–172).

Method 5.8c. Determination of pH in soils with soluble salts

See the discussion under 'pH extractants' in Chapter 4.

Reagents.

- Calcium chloride, 1.0 M – completely dissolve 14.7 g $CaCl_2.2H_2O$ in water and make up to 100 ml.
- Calcium chloride, 0.01 M – completely dissolve 1.47 g $CaCl_2.2H_2O$ in water and make up to 1 l.

Procedure. Proceed as in Method 5.8a., replacing water with a solution of 0.01 M $CaCl_2$. Alternatively, add 5 drops (0.25 ml) of 1 M $CaCl_2$ to the suspension following the pH determination in Method 5.8a.

Calculation. The difference in pH between water and salt solution extracts is known as the *salt effect*, and given the symbol Δ pH. Thus,

$$\Delta \text{ pH} = \text{soil pH in } CaCl_2 \text{ solution} - \text{soil pH in water}$$

and Δ pH values are positive for soils with a net positive charge, and negative for soils with a net negative charge, with magnitude proportional to charge.

Discussion 5.9. Determination of extractable phosphorus

See the discussion under 'Phosphate extractants' in Chapter 4 to determine the most appropriate extractant.

Method 5.9a. Determination of extractable phosphorus (manual method)

The following manual procedure is based on MAFF/ADAS, 1986, pp. 183–185 (with Crown Copyright Permission).

Phosphorus is extracted from soil at 20 ±1°C with a solution of sodium bicarbonate at pH 8.5. The absorbance of the molybdenum blue complex produced by the reduction with ascorbic acid of the phosphomolybdate formed when acid ammonium molybdate reacts with phosphate is measured using a spectrophotometer at 880 nm.

Reagents (extraction).

- Polyacrylamide solution, 0.05% m/v – dissolve 0.5 g of polyacrylamide in approximately 600 ml of water by stirring for several hours. When dissolved, dilute to 1 l.
- Sodium bicarbonate reagent – dissolve 420 g sodium bicarbonate (sodium hydrogen carbonate) in water, add 50 ml of the polyacrylamide solution and dilute to 10 l. Add approximately 50% m/m sodium hydroxide solution, stirring with a glass rod, until the pH meter reading is steady at 8.50 at 20°C (a plastic Pasteur pipette is useful for dropwise addition approaching the required pH).

Procedure (extraction). Transfer 5 ml (scoop filled and struck off level without tapping) of air-dry soil, sieved to ⩽2 mm into a bottle (e.g. wide-mouth, square HDPE). Add 100 ml of sodium bicarbonate reagent, pH 8.50, cap the bottle and shake on a reciprocating shaker, at approximately 275 strokes of 25 mm length per minute, for 30 min at 20°C. Filter a portion immediately through a Whatman No. 2 filter paper, rejecting the first few millilitres of filtrate. Carry out a blank determination.

Reagents (determination).

- Ammonium molybdate reagent, 1.2% m/v – dissolve 24 g powdered ammonium molybdate (ammonium paramolybdate, $(NH_4)6Mo_7O_{24}.4H_2O$), and 0.6 g antimony potassium tartrate (*Note*: cumulative poison) in 1200 ml water. Slowly add 296 ml sulphuric acid (approximately 98% m/m H_2SO_4), stir slowly with a glass rod, and dilute to 2 l. (*Note*: sulphuric acid is highly corrosive and generates heat when diluted; standing the beaker in a sink with a few centimetres depth of cold water before adding the acid will reduce any likelihood of localized boiling. Wear PPE for this step.) Store in a dark glass bottle in a refrigerator.
- Ammonium molybdate reagent, 0.15% m/v – dilute 1 vol. of 1.2% m/v

ammonium molybdate reagent to 8 vol. with water. Store in a dark glass bottle in a refrigerator.

- Ascorbic acid solution, 1.5% m/v – prepare immediately before use, allowing 5 ml per standard, blank and sample, with some spare for any repeats.
- Phosphorus stock standard solution, 1 mg P ml^{-1} – dry potassium dihydrogen orthophosphate at 102°C for 1 h and cool in a desiccator. Dissolve 0.879 g of the dried salt in water and add 1 ml of hydrochloric acid, approximately 36% m/m HCl. Dilute to 200 ml and add 1 drop of toluene to the solution.
- Phosphorus intermediate standard solution, 20 µg ml^{-1} – pipette 10 ml of the phosphorus stock standard solution, 1 mg ml^{-1}, into a 500-ml volumetric flask, make up to the mark and mix. Add 1 drop of toluene to the solution.
- Phosphorus working standard solutions, 0–7 µg P ml^{-1} – prepare fresh daily solutions by pipetting 0, 5, 10, 15, 20, 25, 30 and 35 ml of the phosphorus intermediate standard solution, 20 µg ml^{-1}, into 100-ml volumetric flasks, make up to the mark with sodium bicarbonate reagent, and mix. These will contain 0, 1, 2, 3, 4, 5, 6 and 7 µg P ml^{-1} respectively.
- Sulphuric acid, approximately 1.5 M – slowly with stirring, add 80 ml sulphuric acid, approximately 98% m/m H_2SO_4, to about 800 ml water in a 2-l beaker (*Note*: sulphuric acid is highly corrosive and generates heat when diluted; standing the beaker in a sink with a few centimetres of cold water before adding the acid will reduce any likelihood of localized boiling. Wear PPE for this step.) Cool and dilute to 1 l.

Procedure. Pipette 5 ml of each phosphorus working standard solution (i.e. 0, 5, 10, 15, 20, 25, 30 and 35 µg P) into a plastic vending cup (or 100-ml conical flask). Add 1 ml of approximately 1.5 M sulphuric acid and swirl the solution to assist the release of CO_2. Add 20 ml of 0.15% m/v ammonium molybdate reagent, 5 ml of ascorbic acid solution, 1.5% m/v, swirl to mix and allow to stand for 30 min for colour development. Measure the absorbance in a 10 mm optical cell at 880 nm. The colour is stable for several hours. Construct a graph relating absorbance to µg P. The absorbance values should be approximately 0 to 0.8 for the 0 and 35 µg P standards respectively.

Similarly pipette 5 ml of the soil extract into a plastic vending cup, followed by 5 ml sulphuric acid, 1.5 M. If the soil extract solution is highly coloured after addition of the acid, pipette a duplicate sample and add the 5 ml of sulphuric acid, 1.5 M. Add the other reagents, as detailed above, to the first sample, but only add the ammonium molybdate reagent to the duplicate. Measure the absorbance at 880 nm. For coloured extracts, subtract the absorbance of the duplicate without ascorbic acid, which will not develop the blue colour, from the absorbance of the sample extract with ascorbic acid.

Calculation. Read from the standard graph the number of µg of P equivalent to the absorbances of the sample and blank determinations. Subtract the blank from the sample value, and multiply the difference by 4. The result gives the mg l^{-1} extractable phosphorus in the air-dry soil. This can be expressed for oven-dry soil as in Method 5.2, Calculation (2).

Notes:
1. If one of the standards produces an absorbance that lies significantly away from the standard graph produced by the other standards, or if the whole graph is erratic, repeat as necessary. Detergents containing phosphates should be avoided, but ones such as Decon 90 are phosphate-free.
2. The use of phosphate-free carbon to decolorize soil extracts has been found to give erratic results.
3. A slightly less accurate determination is possible using a colorimeter with a wide bandpass filter, e.g. a simple (non-interference type) purple-red filter but an interference filter in a quality instrument gives results comparable to a spectrophotometer.

Method 5.9b. Determination of extractable phosphorus (automated method)

An automated method for the Lachat QuikChem Automatic Flow Injection Ion Analyzer is given in Missouri Agricultural Experiment Station (1998), pp. 27–29, and is available free of charge from Lachat (Lachat Instruments, 1988). Sun *et al.* (1981) describe a method for the Tecator FIAstar® flow injection system. A method for a segmented continuous flow procedure for both phosphate and potassium was devised by Armitage (1965). The parameters for the phosphate analysis using a dilute HCl soil extractant are outlined below. The manifold diagram (Fig. 5.3) has been modified to allow for the fact that Armitage later changed the Sampler I to a Sampler II module.

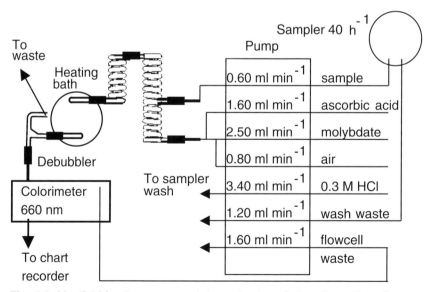

Fig. 5.3. Manifold for the automated determination of phosphorus in soil extracts.

Reagents.

- Hellige–Truog extractant, 0.3 M HCl – dilute 25.8 ml hydrochloric acid, 36% m/m, to 1 l.
- Ammonium molybdate reagent – dissolve 4 g powdered ammonium molybdate in 1 l of water. Slowly with stirring add 80 ml sulphuric acid, 98% m/m H_2SO_4, and dilute to 2 l.
- Ascorbic acid, 0.1% (m/v) – prepare fresh daily.

Procedure. The above reagents, together with the segmenting stream of air, are mixed in a double mixing coil, with the sample being introduced halfway along the coil. A further single mixing coil provides a final mixing before the solution passes through a single glass coil in a thermostatically controlled heating bath at 95°C. After development of the colour, the solution passes to a debubbler, whence it proceeds to the colorimeter fitted with a 15-mm flow-cell, and a pair of 660-nm interference filters. The output is to a chart recorder or personal computer. Standards and blanks should be prepared in the extractant solution according to the normal protocol. It is expected that a sampling rate of 40 h⁻¹, with a time ratio of 2:1 sample:wash would be suitable. This may need adjusting with soils of widely differing extractable P concentrations, where large peaks may obscure very small ones.

Notes:

1. A pump tube has been included for the wash waste for reservoirs that do not have a gravity overflow.
2. The pump tubes in some published manifold diagrams are labelled with the internal diameters of the tubes. We will use the more common convention of labelling with flow rates in ml min⁻¹.
3. Pump tubes are designated by a colour code, which may also be used in the literature. Some common colour codes are listed in Table 5.1.
4. Flow rates are for standard PVC tubing. It is good practice to avoid using the smallest or largest sizes whenever possible.
5. Pump tubes are available in different materials depending on the liquid to be pumped, e.g. for solvents or concentrated acids. Flow rates for these other materials will be different than for standard PVC. Pump tubes may also be available in a specially calibrated or measured flow rate quality at extra cost. Unless specified for medical purposes or to meet regulations, the standard quality is normally adequate. See Chapter 1, 'Peristaltic pumps'.

Method 5.9c. Determination of resin extractable phosphorus (automated method)

The extraction method of Hislop and Cooke (1968), has been outlined in Chapter 4, 'Phosphate extractants'. A blank determination without soil should be carried out. The autoanalysis manifold is shown in Fig. 5.4. Some adjustments to dilution and/or readout sensitivity may be necessary to handle both

Table 5.1. Shoulder colour code for peristaltic pump tubing.

Colour code	Delivery (ml min^{-1})
Orange red	0.03
Orange blue	0.05
Orange green	0.10
Orange yellow	0.16
Orange white	0.23
Black	0.32
Orange	0.42
White	0.60
Red	0.80
Grey	1.00
Yellow	1.20
Yellow blue	1.40
Blue	1.60
Green	2.00
Purple	2.50
Purple black	2.90
Purple orange	3.40
Purple white	3.90

agricultural (lower P) and glasshouse (higher P) soils. The above authors referred to P_2O_5, but we have converted values to P.

Reagents.

- Ascorbic acid, 1% (m/v) – prepare fresh daily.
- Ammonium molybdate – sulphuric acid stock reagent – dissolve 10 g powdered ammonium molybdate in approximately 70 ml water and dilute to 100 ml. Carefully add 150 ml sulphuric acid, 98% m/m H_2SO_4, to 150 ml water in a 600/800 ml beaker while stirring with a glass rod, and allow to cool. Add the molybdate solution with careful stirring and allow to cool.
- Ammonium molybdate – sulphuric acid autoanalysis reagent – dilute 100 ml of the ammonium molybdate – sulphuric acid stock reagent to 1 l with water and mix.
- Phosphorus stock standard solution, 1 mg P ml^{-1} – Dry potassium dihydrogen orthophosphate at 102°C for 1 h and cool in a desiccator. Dissolve 0.879 g of the dried salt in water and add 1 ml of hydrochloric acid, approximately 36% m/m HCl. Dilute to 200 ml and add 1 drop of toluene to the solution.
- Phosphorus intermediate standard solution, 100 µg ml^{-1} – pipette 50 ml of the phosphorus stock standard solution, 1 mg ml^{-1}, into a 500-ml volumetric flask, make up to the mark with sodium sulphate extractant and mix. Add 1 drop of toluene to the solution.
- Phosphorus working standard solutions, 0–35 µg P ml^{-1} – prepare fresh daily solutions by pipetting 0, 5, 10, 15, 20, 25, 30 and 35 ml of the phos-

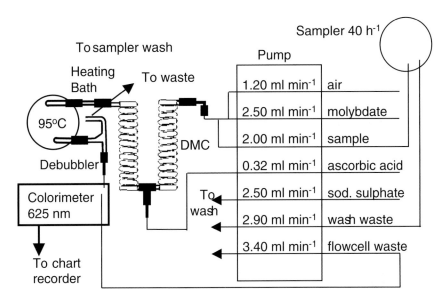

Fig. 5.4. Manifold for the automated determination of phosphorus in soil resin extracts.

phorus intermediate standard solution, 100 µg ml⁻¹, into 100-ml volumetric flasks, make up to the mark with sodium sulphate extractant, and mix. These will contain 0, 5, 10, 15, 20, 25, 30 and 35 µg P ml⁻¹ respectively, and are suitable for glasshouse soils that are approximately ×8 higher in P than agricultural soils; a lower range of 0, 1, 2, 3, 4, and 5 µg P ml⁻¹ should be prepared for the latter.
• Sodium sulphate extractant/wash, 7% (m/v).

Calculation. The 2-ml scoop of soil was extracted via resin into 50 ml sodium sulphate extractant; therefore the concentration must be multiplied by 25 to give the µg P ml⁻¹ in soil by resin extraction. Hislop and Cooke (1968) classified the soils with respect to mg P l⁻¹ air-dry soil as follows:

• agricultural soils: low, <28; medium, 28–65; high, >65
• glasshouse soils: low, <305; medium, 305–436; high, >436.

ADAS have indexed resin P values as follows:

   0, 0–19; *1,* 20–30; *2,* 31–49; *3,* 50–85; *4,* 86–132, *5,* >132 mg P l⁻1.

Method 5.10. Determination of extractable magnesium, potassium and sodium

Magnesium, potassium and sodium are extracted from the soil with 1 M ammonium nitrate.

Reagent (extraction).

- Ammonium nitrate, 1 M – dissolve 400 g of ammonium nitrate in water and make up to 5 l.

Procedure (extraction). Transfer 10 ml (scoop filled and struck off level without tapping) of air-dry soil, sieved to ≤ 2 mm, into a bottle (e.g. wide-mouth, square HDPE), and shake on a reciprocating shaker (approximately 275 strokes of 25 mm per min) for 30 min. Filter through a Whatman No. 2 filter paper, discard the first few millilitres, and retain the rest for analysis of the required elements. Carry out a blank determination.

Reagents (determination).

- Releasing agent – dissolve 13.4 g lanthanum chloride heptahydrate ($LaCl_3$. $7H_2O$) in water and make up to 500 ml.
- Magnesium stock standard solution, 1000 µg Mg^{2+} ml^{-1} – dissolve 1.6581 g magnesium oxide (previously dried at 105°C overnight and cooled in a desiccator) in the minimum of hydrochloric acid (approximately 5 M). Dilute with water to 1 l in a volumetric flask to obtain a solution of 1000 µg Mg^{2+} ml^{-1}.
- Magnesium standards, 10 and 0–1 µg Mg^{2+} ml^{-1}– pipette 5 ml stock solution into a 500-ml volumetric flask and dilute to the mark with M ammonium nitrate reagent to obtain a stock solution of 10 µg Mg^{2+} ml^{-1}. Pipette 0, 2, 4, 6, 8 and 10 ml of the 10 µg Mg^{2+} ml^{-1} stock solution into 100-ml volumetric flasks, add 5 ml releasing agent and make up to the mark with 1 M ammonium nitrate reagent and mix. This will give solutions containing 0, 0.2, 0.4, 0.6, 0.8 and 1.0 µg Mg^{2+} ml^{-1}.
- Potassium stock standard solution, 1 mg ml^{-1} of potassium – dry potassium nitrate at 102°C for 1 h and cool in a desiccator. Dissolve 1.293 g of the dried salt in water and add 1 ml of hydrochloric acid (approximately 36% m/m HCl). Dilute to 500 ml and add 1 drop of toluene.
- Potassium working standard solutions, 0–50 µg ml^{-1} of potassium. Pipette 0, 1, 2, 3, 4 and 5 ml of the potassium stock standard solution into 100-ml volumetric flasks, dilute to the mark with 1 M ammonium nitrate solution and mix. These will contain 0, 10, 20, 30, 40 and 50 µg K ml^{-1}.
- Sodium stock standard solution, 1 mg ml^{-1} of sodium – dry sodium chloride at 105°C for 1 h and cool in a desiccator. Dissolve 0.254 g in water, make up to 100 ml and mix.
- Sodium intermediate standard solution, 20 µg ml^{-1} of sodium – pipette 10 ml of the sodium stock standard solution into a 500-ml volumetric flask, make up to the mark with 1 M ammonium nitrate reagent and mix.
- Sodium working standard solutions, 0–2 µg ml^{-1} of sodium – pipette 0, 2, 4, 6, 8 and 10 ml of the sodium intermediate standard solution into a 100-ml volumetric flask, make up to the mark with 1 M ammonium nitrate solution and mix. These will contain 0, 0.4, 0.8, 1.2, 1.6 and 2.0 µg Na^+ ml^{-1}.

Procedure (determination). Magnesium is determined by atomic absorption spectrophotometry (see Method 5.2, 'Measurement of calcium and magnesium by AAS').

Potassium and sodium are determined by flame photometry (see Method 5.2 'Measurement of potassium and sodium by flame photometry').

Analyse the standards and adjust the zero and maximum standard readings in the usual way.

- Magnesium: pipette 2 ml sample solution into a 100-ml volumetric flask, add 5 ml releasing agent, make up to the mark with 1 M ammonium nitrate and mix. Nebulize into the AAS and record the readings (computer, chart recorder or manually, as appropriate).
- Potassium: nebulize the extract without further dilution.
- Sodium: pipette 10 ml extract into a 100-ml volumetric flask, make up to the mark with 1 M ammonium nitrate reagent and mix.

Calculations.

1. Magnesium. From the standard graph determine the number of μg ml^{-1} of magnesium in the sample, subtract the blank value and multiply the difference by 250 (initial extraction ratio of ×5 multiplied by subsequent ×50 dilution of the extract solution). The result is the number of mg l^{-1} extractable magnesium in the air-dry soil. Include any extra dilution factors, and, if required, convert to oven-dry soil using the appropriate factor, as in Method 5.2, Calculation (2).

2. Potassium. From the standard graph determine the number of μg ml^{-1} of potassium in the sample, subtract the blank value and multiply the difference by 5 (initial extraction ratio). The result is the number of mg l^{-1} extractable potassium in the air-dry soil. Include any extra dilution factors, and, if required, convert to oven-dry soil using the appropriate factor, as in Method 5.2, Calculation (2).

3. Sodium. From the standard graph determine the number of μg ml^{-1} of sodium in the sample, subtract the blank value and multiply the difference by 50 (initial extraction ratio of ×5 multiplied by subsequent ×10 dilution of the extract solution). The result is the number of mg l^{-1} extractable sodium in the air-dry soil. Include any extra dilution factors, and, if required, convert to oven-dry soil using the appropriate factor, as in Method 5.2, Calculation (2).

Method 5.11. Determination of extractable trace elements

For a discussion on the nature of the extractants, see Chapter 4 'Trace element extractants'.

The method described below will use the complexing reagent DTPA (diethylene-triaminepentaacetic acid) to extract, by chelation, copper, iron, manganese and zinc (including zinc on calcareous soils); it also shows promise for monitoring cadmium, nickel and lead in soils receiving sludge applications. The amount of nutrient extracted will vary with extractant pH and concentration, shaking time and temperature, and soil:solution ratio. Keeping these parameters constant will enable valid comparisons with subsequent experiments or advisory tests.

The stock standard solutions may be purchased ready-made for AAS. If made in-house, then the appropriate spectroscopically pure metals, oxides or non-hydrated salts should be used, and oven-dried at 102–105°C for 1 h before weighing. To avoid significant weighing errors, at least 0.2 g of substance should be weighed. Metals and oxides should be dissolved in spectroscopically pure grade acids. At the lower wavelengths used for some of these micronutrients ($\leqslant$250 nm), background absorption from molecular flame species, such as CaO, arising from compounds in the soil extracts can have an interfering effect and cause an elevation in the observed absorption. Some AAS instruments have a background correction facility (e.g. by using the Zeeman effect), and this should be used. An approximate assessment of this effect can be achieved by measuring the absorption with a spectral line close to the one being used, but one not showing an absorption for a dilute solution of that particular element (Slavin, 1968; Christian and Feldman, 1970), while keeping the sensitivity of the instrument the same. Another approach is to make up the standards in a matrix of approximately the same levels of soluble salts as found in the soil extracts. Background interference can be more troublesome with electrothermal than with flame atomizers (Fuller, 1977).

Reagents.

- DTPA extractant – dissolve 3.933 g DTPA in a mixture of 29.844 g TEA (triethanolamine) and 22.22 ml water; stir until dissolved. Add 2.944 g calcium chloride ($CaCl_2.2H_2O$) to 1.1 l of water, and when dissolved, add to the DTPA/TEA solution and make up to about 1.9 l with water. Adjust the pH to 7.3 using hydrochloric acid (approximately 36% m/m HCl) and make up to 2 l.
- Releasing agent – dissolve 2.68 g lanthanum chloride heptahydrate ($LaCl_3.7H_2O$) in water and make up to 100 ml.
- Stock standard solutions, 1 mg ml^{-1} of the metal – purchase or make up as appropriate.
- Working standard solutions – dilute 5 ml of the stock standard solutions to 500 ml with DTPA extractant to give intermediate standards of 10 µg ml^{-1} of the metal. Prepare a range of standards in DTPA extractant for each metal. Suggested values are: cadmium, 0, 0.1, 0.2, 0.3, 0.4 and 0.5 µg Cd ml^{-1}; copper, lead or manganese, 0, 1.0, 2.0, 4.0, 6.0 and 8.0 µg Cu, Pb or Mn ml^{-1}; iron, 0, 2.0, 5.0, 10.0, 15.0 and 20.0 µg Fe ml^{-1}; nickel, 0, 0.2, 0.5, 1.0, 2.0, 3.0 µg Ni ml^{-1}; zinc, 0, 0.5, 1.0, 2.0, 3.0 and 4.0 µg Zn ml^{-1}.

Procedure.
Weigh 10 g air-dry soil, sieved to 2 mm (10 mesh) using a stainless steel sieve into a 175-ml square HDPE (e.g. Nalgene) plastic screw-cap

bottle. Add 20 ml of the DTPA extractant and shake on a reciprocating shaker (275 oscillations of 25 mm length per minute, or similar, but keep constant for all extractions) for 2 h. Carry out a blank extraction. Filter through a Whatman No. 42 filter paper, rejecting the first couple of millilitres, into a polythene hinged-cap sample tube. Pipette 10 ml of filtrate and standards into 25-ml beakers and add 0.5 ml releasing agent to each and mix. They are analysed for the required trace elements by atomic absorption spectroscopy using a suitable range of standards made up in the DTPA extractant. Samples may be diluted with DTPA extractant to reduce excessively high readings to the normal range of the instrument. The wavelengths (nm) of the most sensitive resonance lines for AAS are as follows: Cd, 228.8; Cu, 324.8; Fe, 248.3; Pb, 217.0; Mn, 279.5; Ni, 232.0 and Zn, 213.9.

Calculation. The concentration of trace element ($\mu g \ ml^{-1}$) in the extract is read from the standard curve and the blank reading subtracted; the difference is multiplied by 2 to give the $\mu g \ g^{-1}$ (= $mg \ kg^{-1}$) of the trace element in the air-dry soil. Include any extra dilution factors, and, if required, convert to oven-dry soil using the appropriate factor, as in Method 5.2, Calculation (2).

Discussion 5.12. Determination of extractable sulphur

A helpful discussion of sulphur in soils and its availability to plants is found in Combs *et al.* (1998) and Rowell (1994, pp. 213–215). Plants absorb sulphur mainly in the form of sulphate, which is the main form of sulphur occurring in the soil solution. The SO_4-S is therefore the fraction usually measured. Over 90% of the surface soil sulphur occurs in combination with organic molecules from where it is mineralized to sulphate. The SO_4-S concentration has been found to increase from approximately 5 kg ha^{-1} in the first 30 cm depth of a Wisconsin soil, to approximately 10 kg ha^{-1} in the 30–60 cm profile, and approximately 15 kg ha^{-1} in the 60–90 cm profile. It is therefore recommended that subsoil, as well as topsoil, cores are also taken for analysis. There are other sources of sulphur available to the plant, such as the seasonal effect of precipitation of sulphate-containing rain, especially near industrial areas and conurbations, and the sulphur in applied manure. There is also the sulphur adsorbed by clays and oxides of iron and aluminium, which will increase as the pH decreases below 6.5. The extractant may be water or 10 mM calcium chloride solution, but the latter may displace some adsorbed sulphate. In acidic soils, the available sulphur should include the adsorbed sulphate, therefore calcium phosphate [$Ca(H_2PO_4)_2$] or potassium phosphate (KH_2PO_4), which will extract the adsorbed sulphate, are the extractants of choice. Calcium phosphate is preferred, because the calcium ion depresses the solubility of organic matter to produce a clearer filtrate. This is the method described below. For good reproducibility, it is essential to duplicate the conditions used to form the suspension. These include the temperature and the standing time before measuring the absorbance. A known quantity of sulphate 'seed solution' is usually added to improve the reproducibility of the suspension formation. This not only provides a nucleus to assist the uniform growth of barium sulphate crystals, but also ensures that the solubility product of barium sulphate has been exceeded in the final solution, and thus avoids a concave standard curve in the lower range. Acacia powder (gum acacia, gum arabic) is added to the 'seed solution' to stabilize the $BaSO_4$ precipitate when larger amounts of sulphate are encountered. It may be omitted for soils low in sulphate. The presence of HCl in the seed solution prevents the co-precipitation of barium carbonate, phosphate or hydroxide, which would add to the turbidity.

Method 5.12a. Determination of extractable sulphur (manual method)

Instrumentation. The determination of extracted sulphate-S may be carried out by ICP, ion chromatography or turbidimetry. The ICP procedure measures both organic and inorganic S present in the extract, but has a low methodological error. Ion chromatography may be affected by interference from phosphates and speed of sample throughput, but a suitable method is given by Combs *et al.* (1998). The most widely used method is by turbidimetry, where sulphate is precipitated as a white suspension of barium sulphate by the addition of barium chloride solution to the soil extract. The absorption of light is often measured using a nephelometer, spectrophotometer, or colorimeter at 480 nm, however, measurements at lower wavelengths increase sensitivity, but may incur possible curvature of the calibration graph.

$$BaCl_2 + SO_4^{2-} = BaSO_4\downarrow + 2Cl^-$$

Reagents (extraction).

- Activated charcoal, purified – place about 50 g Darco G-6 activated carbon in a wide-neck screw-cap container, add sufficient calcium phosphate extractant to completely wet it, then cap the bottle and shake for 5 min. Filter slowly with suction through a Buchner funnel, then wash three times successively with deionized water. Test the final leachate with a solution of barium chloride (approximately 1.4% m/v in 0.3 M HCl). If turbidity indicates the presence of sulphate, return the charcoal to a beaker, thoroughly mix with deionized water (boil for 15 min if necessary to get a clear test), refilter, wash and test for S as above. When satisfactory, dry overnight at 105°C and store in a tightly capped bottle.
- Calcium phosphate extractant, 500 mg l^{-1} of phosphorus – dissolve 20.3 g calcium phosphate [$Ca(H_2PO_4).2H_2O$] in water and make up to 10 l.

Procedure (extraction). Weigh 10 g air-dry soil sieved to ⩽2 mm (10 mesh) into a 50-ml conical flask. Add 25 ml of calcium phosphate extractant (50 ml for peat or compost) and shake on a reciprocating shaker (at approximately 200–275 oscillations of 25 mm per minute) for 30 min. If the presence of soluble organic matter is suspected, use a scoop to add 0.15 g purified activated charcoal (carbon), and shake for a further 3 min to enable a subsequent clear filtrate. Filter through a 125-mm Whatman No. 40 (or 42) filter paper, rejecting the first few millilitres. Carry out a blank determination.

Reagents (determination).

- Barium chloride crystals – sieve the powdered $BaCl_2.2H_2O$ crystals, retaining the 520–860 μm (30–20 mesh) fraction. *Warning*: barium chloride is poisonous; wear PPE when handling.
- Seed solution, 20 μg ml⁻¹ of SO_4-S – dissolve 0.1087 g of K_2SO_4 in 500 ml water in a 2-l beaker, and add 500 ml of hydrochloric acid (approximately 36% w/v HCl). Carefully place a Teflon-coated magnetic stirrer-bar into the beaker, place on to a magnetic stirrer and switch on. Add 2 g of acacia powder (see above discussion) slowly while stirring so as to avoid formation of any lumps. Transfer to a bottle and store in a refrigerator. *Safety note*: Acacia powder will cause severe irritation to the eyes, and also irritates the skin and the digestive and respiratory tracts; it is also a mutagen. Wear appropriate PPE in handling acacia powder.
- Sulphate stock standard solution, 500 μg ml⁻¹ of SO_4-S – dissolve 2.717 g potassium sulphate (K_2SO_4), previously dried at 105°C for 1 h and cooled in a desiccator, in calcium phosphate extractant, then transfer to a 1-l volumetric flask with washings and make up to the mark with extractant.
- Sulphate working standard solutions, 0–12 μg ml⁻¹ of SO_4-S – pipette 1, 2, 4, 6, 8 and 12 ml of the 500 μg ml⁻¹ sulphate stock standard solution into 500-ml volumetric flasks, make up to the mark with calcium phosphate extractant and mix. This will give solutions containing 1, 2, 4, 6, 8 and 12 μg ml⁻¹ of sulphate-S.

Procedure (determination). If charcoal has been used in the sample extraction stage, then 25-ml aliquots of the working standards should be shaken with a 0.15 g scoop of purified charcoal for 3 min, and filtered through a 125 mm Whatman No. 40 (or 42) filter paper, rejecting the first few millilitres. Pipette 10-ml aliquots of standards, blank and sample extracts into a 50-ml conical flask containing a magnetic stirrer bar, add 1 ml of the 'seed solution', swirl to mix. Next add a 0.3-g scoop of $BaCl_2.2H_2O$ crystals and stir magnetically for 1 min. Within the next 8 min, read the absorbance at 420 nm on a colorimeter or spectrophotometer fitted with a 40 mm optical cell. The background absorbance resulting from fine clay particles passing through the filter paper should also be measured on the soil extract plus 'seed solution', but without addition of $BaCl_2$ crystals.

Calculation. Read the μg ml⁻¹ SO_4-S for all the solutions from the standard graph. Add the values for the blank and background absorbance, and subtract the sum from the value of the sample solution to give a corrected value. Since 10 g soil provided 25 ml extract, multiply the corrected value of μg ml⁻¹ SO_4-S in the extract by 2.5 to give the μg g⁻¹ SO_4-S (= mg kg⁻¹ SO_4-S) in the air-dry soil. Include any extra dilution factors, and, if required, convert to oven-dry soil using the appropriate factor, as in Method 5.2, Calculation (2).

Method 5.12b. Determination of extractable sulphur (automated method)

An automated method should improve reproducibility by maintaining constant conditions for the formation of the $BaSO_4$ precipitate. A method suitable for the segmented-flow analysis of sulphate in soil and plant extracts using Skalar Analytical equipment has been proposed by Coutinho (1997). Soils are said to be extracted with double-distilled water according to the method in MAFF/ADAS (1986, pp. 215–216), however, that reference uses approximately 1.5 M HCl (10 g soil extracted with 70 ml water and 10 ml HCl, 36% m/m); and the absorbance is measured at 420 nm using a 50-mm cell path. It is suitable for soils up to 10 µg SO_4-S ml^{-1} without a dilution step, and for soils up to 100 µg SO_4-S ml^{-1} with automatic dilution of the sample. A method for soils low in sulphur (up to 1 µg SO_4-S ml^{-1}) using Technicon (Bran and Luebbe) AutoAnalyzer equipment was described by Bettany and Halstead (1972). The Turner Model 111 Fluorometer was modified to enable it to function as a nephelometer, but presumably a colorimeter could be substituted with measurement at 480 nm using a long path (40–50 mm) optical flowcell. The method, with some amendments, is summarized below.

Reagents.

- Barium chloride reagent – add 10.0 g of polyvinyl alcohol, 40.0 g of barium chloride ($BaCl_2.2H_2O$, turbidimetric grade) and 120 ml of 1.0 M HCl to 1800 ml of water. *Warning*: barium chloride is poisonous; wear PPE when handling. Heat on a stirrer-heater unit until the solution clarifies, then allow to cool before adjusting to 2 l. Allow to stand for 2–3 days, then filter through glass wool prior to use.
- Calcium chloride extractant, 0.01 M – completely dissolve 1.47 g $CaCl_2.2H_2O$ in water and make up to 1 l.
- EDTA buffer wash – dissolve 40 g tetrasodium EDTA, and 6.75 g ammonium chloride in approximately 800 ml water, then, in a fume cupboard, carefully add 57 ml ammonia solution (approximately 35% m/m NH_3). Stir and make up to 1 l with water.
- Hydrochloric acid, 1.0 M.
- Hydrochloric acid, 0.5 M.
- Hydrochloric acid, 0.085 M.
- Sodium peroxide solution, 2% m/v. (*Note*: sodium peroxide is highly corrosive, very irritant to skin and mucous membranes, and may ignite combustible materials; wear suitable PPE.)
- Sulphate stock standard solution, 500 µg ml^{-1} of SO_4-S – dissolve 2.717 g potassium sulphate (K_2SO_4), previously dried at 105°C for 1 h and cooled in a desiccator, in water, then transfer to a 1-l volumetric flask with washings and make up to the mark with extractant.

- Sulphate intermediate standard solution, 50 µg ml^{-1} – pipette 10 ml of the sulphate stock standard into a 100-ml volumetric flask, make up to the mark with water and mix.
- Sulphate working standards, 0.25–1.5 µg ml^{-1} – pipette aliquots of 1, 2, 3, 4, 5 and 6 ml of the sulphate intermediate standard solution into 200-ml volumetric flasks, make up to the mark with water and mix. This gives a series of standard solutions of 0.25, 0.50, 0.75, 1.00, 1.25 and 1.50 µg SO$_4$-S ml^{-1}.

Procedure (extraction). Add 25 g air-dry soil, sieved to ≤2 mm (10 mesh), into a 125-ml conical flask. Add 50 ml calcium chloride extractant, 0.01 M, and shake on a reciprocating shaker (at approximately 200–275 oscillations of 25 mm per minute) for 30 min. Filter through a Whatman No. 30 filter paper, rejecting the first few millilitres, and pipette 25 ml of the filtrate into a 50-ml beaker. Slowly add 2 ml sodium peroxide solution using a plastic pasteur pipette. Allow to stand for 5 min, after which the gelatinous precipitate of interfering cations and organic matter is filtered off using a Whatman No. 42 filter paper. Wash the precipitate and filter paper with water, adjust the filtrate to approximately pH 3 with 0.5 M HCl, transfer with rinsing to a 50-ml volumetric flask, make up to the mark with water and mix. Carry out a blank determination.

Procedure (determination). Samples, blank, and standards are poured into the sample cups. Industrial 8.5-ml autoanalyser cups (Part No. 127-0080-01) are available from Gradko or LIP (Equipment & Services) Ltd (see Chapter 1, 'Chemistry module'). Technicon Sampler IV would require the 10-ml cups, same Part No. from Bran + Luebbe. Every 10th cup should contain the EDTA buffer wash solution followed by a cup containing water; this is to prevent build-up of barium sulphate on the walls of the mixing coil and flowcell. Pump the reagents for about 30 min at the start to flex the pump tubes. Sample the highest standard several times and adjust the sensitivity of the colorimeter/spectrophotometer and/or chart recorder to give a reading of about 90% full-scale. Adjust the zero setting and baseline reading to about 5% full-scale for aspiration of the wash solution. If the high standard reading is too low or high, alter the ratio of flow rates of the sample and barium chloride reagent as appropriate. If the readings from the soil samples are too high, take a smaller weight of sample for the extraction. The flow diagram is shown in Fig. 5.5.

Calculation. The 25 g soil was shaken with 50 ml calcium chloride extractant, and 25 ml of this extract was diluted to 50 ml. There is therefore a ×4 dilution factor. Calculate the concentration of SO$_4$-S in the blank and samples by comparison with the standard curve. Subtract the blank value from the sample values and multiply by 4 to give the µg SO$_4$-S g^{-1} (= mg SO$_4$-S kg^{-1}) air-dry soil. Include any extra dilution factors, and, if required, convert to oven-dry soil using the appropriate factor, as in Method 5.2, Calculation (2).

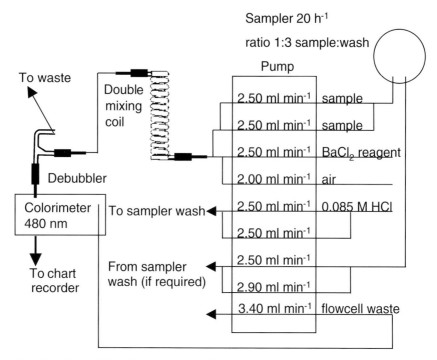

Fig. 5.5. Manifold for the automated determination of sulphate in soil extracts.

The Analysis of Composts

The term *compost* has been defined by Zucconi and Bertoldi (1987) as 'the stabilized and sanitized product of composting which is beneficial to plant growth. It has undergone an initial, rapid stage of decomposition and is in the process of humification.' The initial thermophilic stage of decomposition is the means of self-sanitizing and removing pathogens. If the compost is insufficiently humified, it is immature, and the wide C:N ratio causes it to immobilize soil nitrogen as it continues to actively decompose in the soil. If sufficiently sanitized and humified, the compost is said to be biomature. The development of globally accepted criteria for compost specifications is still at an early stage, so some scientists have proposed biomaturity tests (Mathur *et al.*, 1993).

Methods on-line

The United States Environmental Protection Agency (EPA) has a website where physical and chemical test methods for evaluating solid wastes may be downloaded as pdf files:

http://www.epa.gov/sw-846/main.htm

We understand here by *compost* a marketed product of an organic based material derived from a variety of sources. These might be treated municipal waste, spent mushroom compost, a bracken- or seaweed-based compost, agricultural and food processing wastes etc., which might be put to agricultural use. Composts are often very heterogeneous, which makes it difficult to prepare a sufficiently homogeneous sample. The high humus content makes them similar to peat soils, where organic matter can exceed 95%, which can affect not only the analytical method, but also the interpretation of the results in making fertilizer recommendations.

Typical specifications

Typical parameters and nutrient levels for assessment of compost quality are shown in Table 5.2. These are combined values from a variety of sources, including Bertoldi *et al.* (1987), and are merely intended to help in setting up analytical procedures.

Some typical and preferred heavy and trace element concentrations for soils and municipal composts are shown in Table 5.3. The levels in soils are typical for dilute aqueous extractants such as 0.05 M EDTA, 0.5 M acetic acid, hot water for boron, and, for molybdenum, Tamm's reagent (acid ammonium oxalate; Reisenauer, 1965). Tables in the literature often give total values obtained spectrographically, by XRF, or by extraction with hot

Table 5.2. Typical parameters and nutrient levels for assessment of compost quality.

	Minimum	Normal range	Maximum	Preferred values
Parameter				
Dry matter (DM)		40–60%		lower
Organic matter		20–40% DM		×2 organic C
pH	5.5	6.5–8.0	8.0	7.0–8.0
Salinity (as NaCl)			2.0 g l^{-1}	
(conductivity)		1–2 dS m^{-1}	2.0 dS m^{-1}	≤0.5 dS m^{-1}
Nutrient (g kg^{-1} DM)				
Calcium (CaO)	20			35–140
Calcium (Ca)	14			25–100
Magnesium (MgO)	3	4–16		
Magnesium (Mg)	1.8	2.4–9.7		
Nitrogen (Kjeld.)	6	6–13		High
Organic-N % total	≥90%			
NH$_4$-N		2–5	0.4	Low
NO$_3$-N		50–200		High
NO$_3$-N/ NH$_4$-N		×20–80 ratio		≥60
CN		×10–20 ratio	22	
Phosphorus (P$_2$O$_5$)	5	3.5–14		High
Phosphorus (P)	2.2	1.5–6.0		High
Potassium (K$_2$O)	3	4.5–18		High
Potassium (K)	2.5	3.7–15		High

Table 5.3. Some typical and preferred heavy and trace element concentrations for soils and composts; various local regulations specify different maximum levels.

Heavy and trace elements	Symbol	Typical/ preferred in soil (mg kg^{-1} DM)	Municipal compost (typical) (mg kg^{-1} DM)	Max. in compost (var. regs.)/[a] (mg kg^{-1} DM)	Preferred values (mg kg^{-1} DM)
Boron	B	0.01–10/0.5–<3	0.3–11.9		
Cadmium	Cd	<0.01–1/0	1–8	3–20/1.5	0
Cobalt	Co	<0.05–3/<50	15	34–150	<50
Copper	Cu	0.3–20/1–<3	50–475	100–1200/200	<100
Chromium	Cr	0.1–4.0/low	30–200	150–1200/100	<100
Iron	Fe	<10–3000/10–50	18200		
Lead	Pb	0.1–10/0	65–900	100–1200/150	0
Manganese	Mn	5–100/3–<50	320		
Mercury	Hg	0.02–0.4/0	0.5–4	0.8–5/1	0
Molybdenum	Mo	0.08–0.8/2–7		5–20	
Nickel	Ni	0.2–25	15–100	25–200/50	<25
Zinc	Zn	<2–30/1–8	160–2100	300–3000/400	<300

[a]Upper limit set by The Composting Association (TCA, 2000).

concentrated acids, and these can be about 2–20 times the values obtained with mild extractants. Examples of surveys of trace elements in soils are given by Archer (1980), Archer and Hodgson (1987) and Berrow and Burridge (1980).

Various local regulations specify different maximum levels, so the appropriate authority must be consulted when formulating composts. For a summary of compost standards in Canada, visit the Composting Council of Canada website at:

http://www.compost.org/standard.html

The European Committee for Standardization (CEN) is harmonizing European standards, and it is hoped their draft methods will eventually be introduced as British Standard methods, coded BS EN.

For specifications of a typical commercial compost, visit:

http://www.asiagreen.com.my/chem_ana.htm and /chem_anb.htm

Apart from mandatory specifications, there are voluntary specifications, which permit the compost manufacturer to use the validating scheme's symbol. In the UK, The Composting Association (TCA) have drawn up their Standards for Composts covering certification, testing, monitoring and labelling of composts (TCA, 2000). These may be viewed at:

http://www.compost.org.uk/standard.htm

Specifications for composts for organic farming are seen in the EU Eco-label for Soil Improvers, and The Soil Association Standards for Organic Food and Farming – Certified Products Scheme. Criteria for the award of the EU Eco-label to soil improvers are available at:

http://www.europa.eu.int/comm/environment/ecolabel/soil_improver s/htm

Preparation of sample

Although the cation exchange capacity (CEC) of the compost before application has a poor relationship to its CEC when incorporated into the soil, it provides possibly a better indicator than the C:N ratio of the maturity of the compost (Estrada *et al.*, 1987). It therefore appears as a diagnostic measure of maturity rather than as a parameter defining the quality of the marketed compost. In this context, Jacas *et al.* (1987), have shown that the fineness of grind in the preparation stage can markedly affect the measured CEC. They expressed the results as mEq 100 g^{-1} ash-free sample (CEC/TOM), and the average values using different grinding procedures for four different samples are as follows:

- unmilled sample, 71.3;
- kitchen-grinder milled, 83.2;
- 0.75 mm pulverized, 104.8;
- 0.12 mm pulverized, 111.1.

It is evident that for comparability and consistency, the same procedure should be adopted for subsequent batches, and the proposed method suggests pulverizing to 0.75 mm.

Cation exchange capacity

In the review by Mathur *et al.* (1993) the shortcomings of the CEC method for composts have been highlighted. The CEC of the mineral constituents of compost have been estimated as about 10 mEq 100 g^{-1} DM (Harada and Inoko, 1980), and that of the whole compost at the start, after 5 weeks, and at maturity, as 40, 70 and 80 mEq 100 g^{-1} DM (Harada *et al.*, 1981). Others have found that different mature composts can vary from 27 to 83 mEq 100 g^{-1} DM (for domestic refuse plus sewage sludge, and pine bark plus sewage respectively), and from 72 to 144 for the same samples when expressed as mEq 100 g^{-1} organic matter (Jacas *et al.*, 1987). The assessment of whether compost maturity has been reached is therefore only possible if the initial CEC has been determined for comparison. Jacas *et al.* (1987) describe a simplified unpublished method proposed by Inoko and Harada, an outline of which is given below.

Method 5.13. Determination of CEC in composts

Reagents.

- Hydrochloric acid, 0.05 M.
- Barium acetate reagent, 0.5 M, pH 7.0 – dissolve 127.7 g barium acetate [Ba(CH$_3$COO)$_2$] in water and make up to 1 l. Adjust to pH 7.0 using acetic acid and/or barium hydroxide solution as appropriate.
- Sodium hydroxide, 0.05 M.
- Thymol blue indicator (thymolsulphonphthalein) – triturate 1 g thymol blue

(acid form) in a clean glass mortar with 21.5 ml 0.1 M sodium hydroxide and dilute to 1 l with water.

Procedure. Place 0.2 g compost, milled to 0.75 mm (approximately 24 or 25 mesh) into a sintered glass filter funnel (Porosity 3, 16–40 μm) fitted with a rubber tube with a pinchcock. Using a measuring cylinder or dispenser, add 25 ml hydrochloric acid, 0.05 M, and stir intermittently for 20 min. Open the pinchcock and filter with suction. Repeat the acid treatment, refilter, and wash with successive additions of water (about 150 ml in total) until free of chloride, indicated by the absence of a white precipitate or cloudiness on addition of silver nitrate solution. With the pinchcock closed, add 25 ml 0.5 M barium acetate reagent solution and leave for 1 h. Filter and retain the filtrate, then repeat the treatment with barium acetate reagent, and retain the filtrate. Wash thoroughly with successive additions of water (about 150 ml in total), and combine these washings with the previous filtrates, which are then titrated with 0.05 M NaOH. This may be done potentiometrically to an inflexion point, or using thymol blue (changes from yellow to blue over pH 8.0–9.6). A blank extraction and titration is carried out using the same quantities of barium acetate reagent. The difference in titration figures is equivalent to the H^+ displaced from the sample by the Ba^{2+}, which gives the CEC.

Calculation. If the titre of 0.05 M NaOH is y ml, then:

no. mol $= (y \times 0.05) \times 10^{-3}$, which for monovalent Na^+ is also the no. mol_c.

$(y \times 0.05) \times 10^{-3}$ $mol_c = 5y \times 10^{-5}$ mol_c per 0.2 g air-dry soil, which becomes

$25y \times 10^{-5}$ mol_c g^{-1} air-dry soil, or $25y \times 10^{-2}$ mol_c kg^{-1} air-dry soil, which is

$25y$ $cmol_c$ kg^{-1} air-dry soil (or $25y$ mEq $100g^{-1}$ air-dry soil).

If this is multiplied by 100/total organic matter %, it gives the CEC/TOM value expressed as $cmol_c$ kg^{-1} ash-free air-dry compost. Include any extra dilution factors, and, if required, convert to oven-dry soil using the appropriate factor, as in Method 5.2, Calculation (2).

Other tests adapted from Zucconi and Bertoldi (1987) for certain chemical parameters are listed below:

Ammoniacal nitrogen:	distillation, or ion-selective electrode
Calcium:	extraction with acid followed by AAS or ICP determination
Carbon:	(organic matter) × 0.5, or homogenize plus dichromate oxidation (see Method 5.7b. Procedure: use 0.1 g sample), or elemental analysis
C:N ratio:	calculate from C and N values
Conductivity:	conductivity meter
Heavy metals (total):	Cd, Cu, Cr, Hg, Ni, Pb, Zn are extracted from the milled (≤0.25 mm; 50 mesh Market Grade

	or No. 60 testing sieve) oven-dry sample by acid (e.g. a mixture of HNO_3 and H_2O_2, or HNO_3 and HCl) and determined by AAS or ICP.
Magnesium:	extraction with acid followed by AAS or ICP determination
Moisture:	by difference from value for total solids
Nitrate nitrogen:	(i) by reduction and distillation/titration; (ii) colorimetric determination, or (iii) selective ion electrode
Nitrogen:	(i) by Kjeldahl digest; (ii) elemental analyser, or (iii) dichromate oxidation.
Organic matter:	ignition at 450–600°C
pH:	pH electrode
Phosphorous:	extraction with acid followed by colorimetry
Potassium:	extraction with acid followed by flame photometry
Total solids:	dry at 105°C to constant weight.

Method 5.14. Determination of Ca, K, Mg and P in composts

The above authors did not specify the reagents, however, a suitable acid extractant for Ca, K, Mg and P would be the Mehlich 1 Extracting Solution (Mehlich, 1953), also known as Dilute Double Acid or the North Carolina Extractant. This is a mixed reagent of 0.0125 M H_2SO_4 and 0.05 M HCl, where relatively greater amounts of phosphate are dissolved using the mixed acids than by using HCl alone. This reagent would also dissolve any free carbonates. If necessary, activated carbon (NoritTM SG Extra, or equivalent), which has been purified by washing with extractant, is used to provide colourless extract solutions for the subsequent colorimetric determination of phosphate. Otherwise, the absorbance of solutions, without added reducing agent to develop the colour, are measured and subtracted from the normal absorbance readings (see Method 5.9a).

Reagent.

- Mehlich 1 Extracting Solution, (0.0125 M H_2SO_4 + 0.05 M HCl) – using a measuring cylinder, add 12 ml H_2SO_4 (approximately 98% m/m) and 73 ml HCl (approximately 36% m/m) to 15 l of water in a 20-l polypropylene bottle, dilute to 18 l and mix.

Procedure. Carry out the extraction and then determine phosphorus as in Method 5.9a, except that working standard solutions should be made up in the Mehlich 1 extractant. Determine potassium and magnesium as in Method 5.10, except that standards are made up in Mehlich 1 extractant. Determine calcium as in Method 5.2, except that standards are made up in Mehlich 1 extractant. The composition of composts is so variable that appropriate dilutions of extracts may be required, and calculations adjusted accordingly.

Method 5.15. Determination of heavy metals in compost

To avoid the risk of explosion from using perchloric acid, the following method uses a mixture of nitric and hydrochloric acids (*aqua regia*). This extractant has been satisfactorily used by ADAS for over 10 years. Overnight soaking in cold *aqua regia* reduces frothing on heating. The digestion period has been extended to 3 h, which also makes it suitable for plant material; if there is excess organic matter in the sample, slightly more *aqua regia* is added at the start (Bob Llewelyn, ADAS Wolverhampton, 2001, personal communication). *Note*: TCA recommends extraction method BS EN 13650 with determination by ICP or ISO 11047. Details of behrotest® workstations for *aqua regia* digestion are available at the website:

  http://www.behr-labor.com/ehtemel/prods/01404a.html

Some mercury may be lost by the procedure given below, but excellent results have been reported for the self-contained LECO AMA254 advanced mercury analyser; details from LECO Corporation are at:

  http://www.leco.com/organic/ama/ama254.htm

Reagents (digestion).

- Digest acid – *Safety note*: Wear PPE and use a fume cupboard when handling concentrated acids which emit fumes. Carefully add 390 ml HCl (approximately 36% m/m HCl) to 360 ml of water and stir with a glass rod to mix. Slowly add 250 ml HNO_3 (approximately 70% m/m HNO_3) and stir slowly to mix, avoiding splashes. It is recommended that extra pure, spectroscopy (AAS/ICP) grade acids are used for trace element analyses.
- Nitric acid–KCl reagent, 8.75% m/m HNO_3 – carefully add 125 ml HNO_3 (approximately 70% m/m HNO_3) to 40 ml of KCl solution (see below) and dilute with water to 1 l and mix.
- Potassium chloride solution, 5% m/v – dissolve 5 g of KCl in water and make up to 100 ml.

Procedure (extraction). Weigh 1.2 g air-dry soil, sieved to ≤2 mm in a stainless steel sieve and then ground in a pestle and mortar, into a borosilicate boiling tube that has a graduation at 60 ml. Add 15 ml of the digest acid and carefully swirl to wet the sample thoroughly. Place a small glass funnel into the neck of the tube (to permit refluxing during the heating stage) and allow to stand overnight. Place the tube in a thermostatically controlled aluminium heating block and raise the temperature to 50°C and maintain this temperature for 30 min. Raise the temperature to 120°C, and digest for 3 h. Allow to cool, remove the funnel carefully to avoid acid drips, and make up to the 60 ml mark with the nitric acid–KCl reagent. Filter through a Whatman 541 paper, discarding the first few millilitres, and retain for analysis. Carry out a blank determination omitting the soil.

Reagents (determination).

- Standard solutions – make up standard solutions, diluting with nitric

acid–KCl reagent, to cover the following concentration ranges in µg element ml^{-1}: Cd, 0–0.2; Cu, 0–2; Pb, 0–2; Ni, 0–0.5 and Zn, 0–1.5.
- Releasing agent – dissolve 2.68 g lanthanum chloride heptahydrate (LaCl$_3$.7H$_2$O) in water and dilute to 100 ml.

Procedure (determination). Pipette 20 ml of each standard, extract and the blank solution into 50 ml beakers and add 1 ml releasing agent, then swirl to mix. Determine using AAS or ICP. If necessary, dilute a fresh sample extract or standard solution with the appropriate amount of nitric acid–KCl reagent to bring the readings on scale, then add releasing agent as above. The most sensitive resonance lines for AAS are (nm): Cd, 228.8; Cu, 324.8; Ni, 232.0; Pb, 217.0 and Zn, 213.9.

Calculation. If the measured concentration of trace element was y µg ml^{-1}, this becomes $y \times 10^{-3}$ mg ml^{-1}. A sample of 1.2 g soil was extracted into 60 ml solution, therefore 1 kg is equivalent to $10^3(60/1.2)$ ml. Therefore the measured concentration is equivalent to $y \times 10^{-3} \times 10^3(60/1.2)$ mg kg^{-1}, which becomes 50y mg kg^{-1} air-dry soil.

6   The Analysis of Fertilizers

There are several publications detailing standard or officially recognized methods of fertilizer analysis. These include *Official Methods of Analysis of AOAC International* (Horwitz, 2000); *Official and Standardised Methods of Analysis*, published by the Royal Society of Chemistry (Watson, 1994); and *Fertilisers – Methods of Analysis used in OEEC Countries* (OEEC, 1952). There are also the EEC methods, which have been implemented in the UK by the *Fertilisers (Sampling and Analysis) Regulations 1996* (Statutory Instrument (SI) 1996 No. 1342). The title page of the SI may be downloaded from the following website:

> http://www.hmso.gov.uk/si/si1996/Uksi_19961342_en_1.htm

with the Schedule 2, Methods of Analysis at:

> http://www.hmso.gov.uk/si/ si1996/Uksi_19961342_en_4.htm#sdiv2

Note: By changing the year and number of the SI using the above URL format, it is possible to access and download any available SI. The methods detailed in the above SI are used by Public Analysts to determine whether a fertilizer conforms to the Fertiliser Regulations 1991 (and subsequent amendments), which is available at the website:

> http://www.hmso.gov.uk/si/si1991/Uksi_19912197_en_1.htm

The British Standards Institute has published their BS fertilizer analysis methods, and these are listed in pages 728–729 of Watson (1994), e.g. nitrate nitrogen is method BS 5551.

The European Commission Directive 77/535/EEC of 22 June 1977 on the approximation of the laws of the Member States relating to methods of sampling and analysis for fertilizers is obtainable at:

http://europa.eu.int/eur-lex/en/lif/dat/1977/en_377L0535.html
where there are links to subsequent amendments from 1979 to 1995.

The Government of India Fertilizer (Control) Order 1985, Schedule-I, Specifications of Fertilizers, is downloadable from:

http://agri.mah.nic.in/agri/input/html/fert_cont_schedule_I.htm
and Schedule II, Part A, dealing with sampling, and Part B, Method of Analysis of Fertilizers, is available at:

http://agri.mah.nic.in/agri/input/html/fert_cont_schedule_II.htm
The methods of analysis are mainly based on those of the AOAC (1965 edition) and the National Plant Food Institute, Washington, DC, 1961.

Methods of fertilizer analysis are also occasionally reviewed in the Proceedings of the International Fertiliser Society, and the contents of all the Proceedings are viewable at:

http://www.fertiliser-society.org/Proceedings/ProcMenu.htm
The following is only a selection of analytical methods because there would be too many to include in a handbook dealing with many substances besides fertilizers. However, the references to articles, and the websites of official methods will provide an extra resource. *Warning!* – It should be emphasized that there are occasional misprints in official methodology (e.g. in the EU method for citric acid extractable phosphate in fertilizers, many full-stops appear as '7', so the citric acid monohydrate appears as having $7H_2O$); only purchased printed versions are actually 'official'. The methods below should give acceptable results in an educational/research context, but may not be as rigorous as the official procedures; the latter should be followed in all cases where ensuing litigation may be a possibility.

Fertilizer Analytical Procedures

Discussion 6.1. Determination of total nitrogen in presence of nitrate and organic N

Nitrogen may be present in several forms: ammonium, cyanamide (NH_2CN), nitrate, urea (carbamide, $CO(NH_2)_2$), and slow-release ureaformaldehyde condensates. Any nitrate must first be reduced to ammonia/ammonium by the use of an appropriate reducing agent. Some reducing agents are:

- Arnd's alloy (60% Cu, 40% Mg) in 20% (m/v) $MgCl_2.6H_2O$ solution containing 1.5% (m/v) $MgSO_4.7H_2O$ and 0.2% (m/v) MgO.
- Devarda's alloy (45% Al, 5% Zn, 50% Cu) in 30% (m/v) NaOH solution.
- Raney catalyst powder (50% Ni, 50% Al) in 20% v/v H_2SO_4 containing 10.67% (m/v) K_2SO_4.
- Ulsch method, powdered iron, reduced in hydrogen, in 1.6% (m/v) H_2SO_4.
- Zinc dust and salicylic acid (5% m/v) in H_2SO_4 (93–98% m/v).

Calcium cyanamide and urea must be subjected to a Kjeldahl digest to convert —NH_2 to NH_4^+. Various modifications to procedures are necessary when several different compounds are present in the same fertilizer sample, and these may be consulted in the above references. It is important to avoid generating heat in the sample grinding process, which could affect any heat-sensitive organic constituents or cause moisture loss; a pestle and mortar is therefore recommended. Conversely, the absorption of atmospheric moisture by hygroscopic constituents, such as ammonium nitrate, must be avoided; therefore the sample preparation stage should be accomplished as speedily as possible before storing in an inert airtight container. The sample is finely ground for two reasons: the fertilizer may contain various compounds with different crystal properties which could lead them to segregate, causing a loss in homogeneity; second, only a semi-micro amount of sample is taken, therefore it is vital that it should be homogeneous and truly representative of the bulk sample. The following procedure uses salicylic acid and zinc dust as reducing agents, and the reactions involved are described in Method 5.6a.i, Reduction of nitrate prior to digestion and colorimetric autoanalysis. The digestion is carried out using micro-Kjeldahl digestion units to benefit from the various economies of scale. Potassium sulphate is omitted from the digestion mixture so that the same digest solution may be analysed for potassium; however, because of the resulting lower boiling point, the digestion may take 4–5 h.

Method 6.1a. Determination of total nitrogen in presence of nitrate and organic N, with final determination by distillation

Reagents.

- Salicylic acid, 2-HOC$_6$H$_4$COOH
- Selenium powder
- Sulphuric acid, approximately 98% m/m H$_2$SO$_4$
- Zinc dust

Apparatus.

- Micro-Kjeldahl digestion unit
- Kjeldahl digestion flasks, 50 ml
- Distillation unit – the Markham semi-micro distillation unit is suitable, or a proprietary automatic unit.

Procedure. Take a representative of the bulk fertilizer sample and reduce in size by cone and quartering, or use a sample divider. Then grind and sieve to 0.2 mm (No. 70 or 70 mesh), mix thoroughly and immediately transfer to an airtight container. Weigh 0.250 g (in duplicate) into a 50-ml micro-Kjeldahl flask, add 25 ml sulphuric acid (approximately 98% m/m H$_2$SO$_4$), 0.5 g salicylic acid (free of lumps), swirl until dissolved and allow the amber coloured solution to stand for 30 min. Add 0.5 g zinc dust, swirl to mix, and allow to stand until evolution of gas subsides. Add 0.1 g selenium powder and 2 or 3 granules of carborundum (pumice-stone is not suitable because it floats in sulphuric acid) to avoid bumping occurring on boiling. Boil gently on a micro-

Kjeldahl heating rack until clear or a pale straw colour, then for a further 30 min. Allow to cool, and then add slowly with stirring to about 100 ml water in a 250-ml beaker. Dissolve the needle-like crystals remaining in the flask and rinse well with water, add the washings to the beaker and allow to cool. Transfer to a 250-ml volumetric flask with beaker washings, and make up to the mark with water and mix. Carry out a blank digestion omitting the sample (if the reduction step is omitted because of absence of nitrate, the sample should be replaced with 0.25 g sucrose to ensure partial reduction of any nitrate in the reagents). The total N, now present in the sample solution as ammonium sulphate, can be determined on a 10-ml aliquot by the distillation procedure given in Method 5.6b., Determination of organic plus ammonium-N by digestion and distillation.

Calculation. Subtract the blank titre from the sample titre, multiply the difference by 7 and divide by the sample weight in grams. This gives the g kg^{-1} of nitrogen in the fertilizer sample. Check duplicate values are sufficiently concordant, and take the mean value. If not, repeat the distillations; if still not concordant, repeat the digestion, ensuring that the original sample has been adequately ground and mixed before weighing.

   Example: ammonium nitrate – sample weight, 0.25 g (which contains 43.8 mg NO_3-N); sample titre, 12.69 ml; blank titre, 0.20 ml of 0.01 M H_2SO_4. The calculation as given above is: $(12.70 – 0.20) \times 7/0.25 = 12.50 \times 28 = 350.00$ g kg^{-1} N in ammonium nitrate. This may be checked as follows: molar mass of $NH_4NO_3 = 80.04$ g; relative formula mass of $2 \times N$ is 28.02 g. Thus the total weight of nitrogen in 1 kg ammonium nitrate is: $10^3 \times (28.02/80.04) = 350.07$ g kg^{-1}.

Method 6.1b. Determination of total nitrogen in presence of nitrate and organic N, with final determination by autoanalysis

Reagents.

- Salicylic acid, 2-HOC_6H_4COOH
- Selenium powder
- Sulphuric acid, approximately 98% m/m H_2SO_4
- Zinc dust

Apparatus.

- Micro-Kjeldahl digestion unit
- Kjeldahl digestion flasks, 50 ml, scratched with a graduation line at 25 ml
- Autoanalysis equipment

Procedure. Take a representative of the bulk fertilizer sample and reduce in size by cone and quartering, or use a sample divider. Then grind and sieve to 0.2 mm (No. 70 or 70 mesh), mix thoroughly and immediately transfer to an airtight container. Weigh 0.0500 g (in duplicate) into a 50-ml micro-Kjeldahl

flask, add 25 ml sulphuric acid (approximately 98% m/m H_2SO_4), 0.1 g salicylic acid, swirl until dissolved and allow the amber coloured solution to stand
for 30 min. Add 0.1 g zinc dust, swirl to mix, and allow to stand until evolution of gas subsides. Add 0.02 g selenium powder and 2 or 3 granules of carborundum (pumice-stone is not suitable because it floats in sulphuric acid) to
avoid bumping occurring on boiling. Boil gently on a micro-Kjeldahl heating
rack until clear or a pale straw colour (approximately 3–4 h), then for a further 30 min. Allow to cool, adjust to the 25 ml scratch line with sulphuric
acid, then add slowly with stirring to about 15 ml water in a 100-ml beaker,
and allow to cool. Rinse the flask into the beaker with a few millilitres of water,
then transfer to a 50-ml volumetric flask with washings and make up to the
mark with water and mix. Carry out a blank digestion omitting the sample.
The total N, now present in the sample solution as ammonium sulphate, can
be determined by autoanalysis; see Chapter 7 for details of standards, reagents
and method.

Calculation. The concentration in mg N l^{-1} read from the standard curve, after
subtracting the blank, is equal to the g kg^{-1} of nitrogen in the sample. Take
the mean of duplicates; if they differ by more than the acceptable experimental error, repeat the autoanalysis, and, if still unacceptable, repeat the
digestion procedure, ensuring adequate grinding and mixing to obtain a homogeneous sample powder, which is important with such a small sample weight.

Example: ammonium nitrate – sample weight, 0.05 g. The blank and sample peaks are compared with the standard curve in the normal way. The blank
peak value (15 mg N l^{-1}) is subtracted from the sample peak value (365 mg
N l^{-1}) to give 350 mg N l^{-1} in the 50 ml of diluted sample digest solution.
The weight of N in 0.05 g sample is therefore:

$(350/1000) \times 50/1000$ g,

thus

the weight of N per kg = $(350/1000) \times (50/1000) \times (1000/0.05) = 350$ g N kg^{-1}.

Discussion 6.2. Determination of phosphorus in fertilizers

The analysis of phosphorus in fertilizers can be achieved in many ways, and
due regard must be made both to the chemical form in which the phosphorus occurs, and its solubility, and hence relative rate of availability to the
plant. Phosphorus can occur in many different molecular combinations and
admixed with a variety of other substances, organic and inorganic, so that it
is difficult to recommend one method in preference to any other. The type
and amount of accompanying trace elements can also vary. The determination of total phosphorus is easier to achieve than available phosphorus, but
the acid digest should not be highly coloured or else it could result in too
high a colorimetric reading.

Phosphate rocks

Phosphate rocks usually contain apatite, and this is mainly virtually insoluble calcium fluorapatite. The molecular formula can be written either as $3Ca_3(PO_4)_2.CaF_2$ or $9CaO.3P_2O_5.CaF_2$. It has been formed from partial or total replacement of the hydroxyl in hydroxyapatite, $3Ca_3(PO_4)_2.Ca(OH)_2$, by fluoride. In addition, varying amounts of hydroxyl may also be replaced by carbonate, or chlorine. Thus there is also carbonapatite, $3Ca_3(PO_4)_2.CaCO_3$; and chlorapatite, $3Ca_3(PO_4)_2.CaCl_2$. Phosphate rocks with a high CO_2 content generally have a very small grain size and high surface area, which facilitates their solubility. They are known as 'soft-earth' or reactive phosphate rocks (RPRs), such as Tunisian Gafsa phosphate. These have been formed on the sea floor in a different geochemical process to the hard variety. Their slow solubility is enhanced by acid soils, high rainfall and a warm climate. It is the presence of the carbonate that enables the finely ground fertilizer to dissolve slowly in the acid soil. Further details on RPRs may be found at the following websites:

> http://www.latrobe.edu.au/www/rpr/what.htm
> http://www.fertico.com.au/rprbrochure.html

The elemental composition of various formulations is available at:

> http://www.fertico.com.au/rock.htm

Another naturally occurring phosphate mineral is sombrerite (whitlockite), which is tricalcium phosphate, $Ca_3(PO_4)_2$.

Superphosphates

The 'hard-earth' coarsely crystalline forms of insoluble fluorapatite are virtually insoluble if the pH is above 5.5. Their very slow release rates mean that about four times the recommended rates for more soluble forms of phosphorus needs to be applied to correct an immediate deficiency. For the phosphate to be immediately available to plants, they need to be treated with acid to convert them to the soluble superphosphate. Sulphuric acid results in single superphosphate, and phosphoric acid produces triple superphosphate. In about 1840, Liebig proposed treating bone phosphate with sulphuric acid to increase the availability of P to plants. However, the conversion of insoluble hard rock phosphate to soluble superphosphate by treatment with sulphuric acid was patented in 1842 by Sir John Bennet Lawes, who opened the world's first artificial fertilizer factory the same year. He also founded Rothamsted Experimental Station, Harpenden, UK, the world's oldest agricultural research station still in existence. The chemical reaction of rock phosphate with sulphuric acid first forms free phosphoric acid, which reacts with more rock phosphate to give calcium dihydrogen phosphate (acid phosphate; monocalcium phosphate) and gypsum (anhydrite), together called single superphosphate (approximately 20% P_2O_5). Fluorine is removed as hydrogen fluoride, and the reaction equations are:

$$CaF_2 + H_2SO_4 = CaSO_4 + 2HF\uparrow$$

$$Ca_3(PO_4)_2 + 2H_2SO_4 = Ca(H_2PO_4)_2 + 2CaSO_4$$

Thus one formula Ca is converted to dihydrogen phosphate, and the product is called single superphosphate (the *super* refers to its solubility/availability compared with rock phosphate).

With phosphoric acid, the main equation becomes:

$$Ca_3(PO_4)_2 + 4H_3PO_4 = 3Ca (H_2PO_4)_2$$

Thus all three formula Ca atoms are converted to dihydrogen phosphate, and the product is therefore called triple superphosphate (approximately 46% P_2O_5 or 20% P).

Basic slags

Slags are usually obtained as a by-product from steel making, where phosphate is removed from the steel to prevent the level rising above 2% P when it becomes brittle. The process is carried out in a Thomas converter by adding lime and silicate to the molten metal, and blowing through it air previously heated to 1600°C. This oxidizes the phosphorus to calcium silicophosphate $((Ca_3(PO_4)_2.x(Ca_2SiO_4))$, which is called Thomas slag. This is then finely ground. They can contain varying amounts of phosphorus, but are preferred to contain at least 5% total P (11.5% P_2O_5) of which not less than 80% is soluble in 2% citric acid. Cadmium can sometimes cause a toxicity problem. Further details are viewable at:

http://soils-earth.massey.ac.nz/cybsoil/article/slag.htm

Organic phosphorus

A common source of organic phosphorus is bone meal (approximately 9–14% P) and bone ash (approximately 18%). The bird excrement guano contains about 2–3% P as ammonium and calcium phosphates. Fresh solid dairy cattle manure has approximately 0.13% P (moisture = 81.7%), and solid swine manure has about 0.33% P (moisture = 71.8%), which will be in both organic and mineral forms.

Solubility of phosphates

Phosphatic fertilizers usually contain a mixture of phosphates exhibiting varying degrees of solubility, which also depend on the nature of the soil. It is therefore necessary to analyse for these various phosphate types. A list of various phosphate compounds, their molecular formulae and solubilities, where known, is given in Table 6.1.

The methods of analysis for phosphate solubility are not absolute, but empirical; that is, they are based on practical experience. For example, the neutral ammonium citrate method is favoured in the USA because it has received over 100 years of study and experimentation, and provides an index correlating the laboratory results with the fertilizing value of water-insoluble phosphates under the conditions prevailing in the principal farming regions of the country. The particular solvent is therefore not an attempt to accurately reproduce the properties of the soil solution in the immediate vicinity of the

Table 6.1. Chemical forms of phosphate with their solubilities, where known, in various solvents.

			Solubility in solvent			
Name	Formula	Water	Neutral amm. citrate (Fresenius)	Alkaline amm. citrate (Petermann) or Joulie	Citric acid 2% (Wagner)	Formic acid 2%
Monocalcium phosphate (super/triplephosphate)	Ca(H$_2$PO$_4$)$_2$	High	High			
Dicalcium phosphate	CaHPO$_4$	Low	High			
Tricalcium phosphate	Ca$_3$(PO$_4$)$_2$.2/3 H$_2$O	V.low	Low			
Tetracalcium phosphate	Ca$_4$P$_2$O$_9$		High			
Apatite (fluorapatite)	3Ca$_3$(PO$_4$)$_2$.CaF$_2$		Low			
Basic slag (Thomas phosphates)	Ca$_3$(PO$_4$)$_2$.x(Ca$_2$SiO$_4$)	Insol.			High	
Rhenania phosphate	3CaNaPO$_4$.Ca$_2$SiO$_4$	Insol.		High		
Soft earth phosphate	3Ca$_3$(PO$_4$)$_2$.CaCO$_3$	V.low				High

plant roots, but to provide a standard measure of phosphate extraction roughly comparable to that available to the plant, which can be related to the observable plant growth. It should also be remembered that the continuing solubility of sparingly soluble calcium phosphates depends on the removal of the solution of the calcium and phosphate ions from the region around the fertilizer granule as soon as they have been formed. With certain tropical soils, the presence of significant amounts of aluminium and iron can fix the phosphate in an unavailable form. A further discussion of the use and analysis of phosphates is found in Sauchelli (1965); there are also useful details on the solubility of phosphate fertilizers in Finck (1982).

Superphosphate contains a mixture of monocalcium phosphate (soluble in water), plus dicalcium phosphate and calcium sulphate (both with low solubility in water) together with other mineral residues. Thus the solubility in water gives the monocalcium phosphate (sometimes abbreviated to monophosphate) content, and extraction of the residue with neutral ammonium citrate gives the dicalcium phosphate component.

Method 6.2a. Determination of water-soluble phosphorus (extraction)

Procedure. The AOAC and the SI 1996 No. 1342 methods differ, and both are outlined below.

1. AOAC (Method 977.01 Preparation of solution) extraction: Weigh 1.000 g of the ground sample and transfer into a 9-cm filter paper in a funnel. Wash with successive small portions of water by directing a jet of water from a wash bottle around the entire periphery of the paper and into the residue, which should be well mixed up with each washing, and drain through completely before the next addition. Continue until the filtrate amounts to about 250 ml. Use suction only if the process would take longer than 1 h. Any turbidity should be removed by the addition of 1–2 ml HNO_3 to the filtrate. Make up to 250 ml with water and mix. A sample of triple superphosphate should give a solution of approximately 800 µg P ml^{-1}.

2. SI 1996 No. 1342 extraction: Weigh 5 g prepared sample to the nearest 0.001 g, and place in a 500-ml volumetric flask. Add 450 ml water at 20–25°C and shake for 30 min on a rotary shaker at 35–40 turns per minute. Make up to the mark with water and mix. Filter through a dry fluted filter paper into a dry container. A sample of triple superphosphate should give a solution of approximately 2000 µg P ml^{-1}.

Method 6.2a.i. Determination of water-soluble phosphorus (autoanalysis)

Procedure. Pipette 1 ml of the AOAC extract (increase appropriately if the peaks are too small) into a 20-ml volumetric flask. Using a pipettor, carefully add 1 ml H_2SO_4 (approximately 98% m/m), swirl slowly to mix and allow to cool. Make up to the mark with 50% (v/v) H_2SO_4 (approximately 98% m/m), stopper and invert several times to mix. The water soluble P, present in the sample solution as orthophosphate, can be determined by autoanalysis; see Chapter 7 for details of standards, reagents and method.

For the SI 1996 No. 1342 extract, pipette 1 ml of the extract (increase appropriately if the peaks are too small) into a 50-ml volumetric flask. Using a pipettor, carefully add 1 ml H_2SO_4 (approximately 98% m/m), swirl slowly to mix and allow to cool. Make up to the mark with 50% (v/v) H_2SO_4 (approximately 98% m/m), stopper and invert several times to mix. The water soluble P, present in the sample solution as orthophosphate, can be determined by autoanalysis; see Chapter 7 for details of standards, reagents and method.

Calculation. For both the AOAC and SI 1996 No. 1342 extracts, divide the concentration of phosphate-P in the diluted sample (read from the standard curve) by 2 to obtain the percentage water-soluble P in the sample. If 2 ml (instead of 1 ml) extract were diluted to 20 ml, divide by 4, and so on. If required, multiply the % P by 2.2915 to obtain the % P_2O_5.

Method 6.2a.ii. Determination of water-soluble phosphorus (manual method)

This method and the following (6.2b.) are based on the methods given by Craven and Schwer (1960). The concentrations of standard solutions are expressed in terms of mg P_2O_5, sometimes incorrectly termed phosphoric acid. We have retained this format in order to give exact increments of concentration for the sequence of standards. It is now conventional, however, for the analytical method to express phosphorus concentration in terms of elemental P, and then convert to % P_2O_5, which is still used for the labelling of fertilizers.

Reagents.

- Nitric acid, approximately 70% m/m HNO_3
- Phosphorus stock standard solution, 1000 µg P_2O_5 ml^{-1} (436 µg P ml^{-1}) – dissolve 1.9173 g potassium dihydrogen orthophosphate, previously dried for 1 h at 105°C, in water. Transfer with beaker washings to a 1-l volumetric flask, make up to the mark and mix.
- Phosphorus intermediate standard solution, 200 µg P_2O_5 ml^{-1} (87.3 µg P ml^{-1}) – dilute 20 ml of the phosphorus stock standard solution to 100 ml with water.
- Potassium dihydrogen orthophosphate
- Sodium hydroxide solution, 1 M
- Vanadium molybdate reagent – *Note*: wear PPE to prevent injury from concentrated nitric acid. Separately dissolve 20 g of ammonium molybdate and 1 g of ammonium vanadate in water, transfer to a 250-ml beaker and swirl to mix. Slowly add 140 ml nitric acid (approximately 70% m/m HNO_3), carefully transfer to a 1-l volumetric flask, make up to the mark with water and mix.

Procedure (standard curve). Fill a 50-ml burette with the phosphorus intermediate standard solution and dispense 25, 26, 27, 28, 29, 30 and 31 ml into a series of 100-ml volumetric flasks. These will contain 50, 52, 54, 56, 58, 60 and 62 µg P_2O_5 ml^{-1} (21.8, 22.7, 23.6, 24.4, 25.3, 26.2, and 27.1 µg P ml^{-1}). Add 25 ml of the vanadium molybdate reagent and make up to the mark with water (both liquids at 20°C), mix and stand for 10 min.

Using a matched pair of cells, place the lowest standard in the reference beam, and measure the absorbance of the other standards at 420 nm. Plot the standard curve relating absorbance to known concentration.

Procedure (sample analysis). Dilute an aliquot of the extract from Method 6.2a. to give ($\geqslant$ 25 ml) a solution containing 220–248 µg P_2O_5 ml^{-1} (96–108 µg P ml^{-1}) at 20°C. Thus for super triplephosphate by the AOAC extraction, dilute 13 ml extract to 100 ml; and for the SI 1996 No. 1342 method, dilute 5.2 ml extract to 100 ml. Pipette 25 ml of this solution into a 100-ml volumetric flask, add 25 ml vanadium molybdate (20°C), make up to the mark with water (20°C), mix and stand for 10 min. Simultaneously prepare a fresh 50 µg P_2O_5 ml^{-1} (21.8 µg P ml^{-1}) reference standard against which the absorbance of the samples are measured.

Calculation. Divide the concentration of the sample solution read from the standard curve by 3 (if diluted as in above sample analysis procedure), to obtain the % P, or divide by 1.304 to obtain the % P_2O_5 in the sample.

Explanation for the AOAC methodology extract: if the diluted sample concentration as read from the standard curve was 60 µg P_2O_5 ml^{-1}, this corresponds to a concentration of 60 × 100/13 × 100/25 = 1846 µg P_2O_5 ml^{-1} = 1846 × 0.4364 = 805.7 µg P in the original extract (because 13 ml was diluted to 100 ml, and 25 ml of this solution was taken and diluted to 100 ml). There was 250 ml of original extract, therefore this volume would contain: 250 × 1846 µg P_2O_5 = 461,500/10^6 g P_2O_5 = 0.4615 g per 1.0 g sample, or 46.15% P_2O_5 in the original sample.

This is equivalent to 46.15 × 0.4364 = 20.14% P. Since 60 µg P_2O_5 ml^{-1} in the final sample solution corresponds to 46.15% P_2O_5 in the original sample, sample concentrations from the standard curve should be multiplied by 46.15/60 = 0.769 to give the % P_2O_5, or 0.336 to give the % P in the original sample.

For the SI 1996 No. 1342 extract, the condensed calculation is: 60 × 100/5.2 × 100/25 = 4615 µg P_2O_5 ml^{-1} = 2014 µg P in the original extract. There was 500 ml of original extract, therefore this volume would contain: 4615 × 500 µg P_2O_5 = 2.3075 g P_2O_5 per 5 g sample, or 46.15% P_2O_5, equivalent to 20.14% P in the original sample. Thus the same factor applies as with the AOAC extract, and sample concentrations from the standard curve should be multiplied by 0.769 to give the % P_2O_5, or 0.336 to give the % P in the original sample.

Method 6.2b. Determination of 2% citric acid-soluble phosphorus – method for basic slags (Thomas phosphate)

Reagents.

- Citric acid monohydrate, crystallized [HOC.COOH(CH$_2$COOH)$_2$.H$_2$O]
- Citric acid extractant – dissolve 10 g citric acid monohydrate in water, make up to 500 ml and adjust to 20°C. This is sufficient for one sample – for a number of samples, increase quantities as appropriate. The concentration may be checked by titrating 10 ml reagent against 0.1 M NaOH using phenolphthalein indicator, when the titre should be 28.55 ml.

Procedure (extraction). Weigh 5.000 g of the finely powdered basic slag sample into a weighing funnel and transfer to a 1-l stoppered bottle. Add 500 ml citric acid extractant at 20°C to the sample while shaking the bottle to avoid caking of the sample. Shake for 30 min; a rotary shaker should be set to 35–40 turns per min. Filter immediately through a dry fluted Whatman No. 4 filter paper; discard the first 20 ml, and collect a sufficient quantity for the analysis in a dry glass receiver. If the basic slag contained 9.2% P_2O_5 (4% P), the extract will contain 920 µg P_2O_5 ml^{-1} (401.5 µg P ml^{-1}).

Procedure (standard curve). Fill a 50-ml burette with the phosphorus intermediate standard solution and dispense 25, 26, 27, 28, 29, 30 and 31 ml into a series of 100-ml volumetric flasks. These will contain 50, 52, 54, 56,

58, 60 and 62 µg P_2O_5 ml^{-1} (21.8, 22.7, 23.6, 24.4, 25.3, 26.2 and 27.1 µg P ml^{-1}). Add 25/y ml of 2% citric acid extractant to compensate for that added in the 25 ml sample extract solution, where y is the dilution factor. In the example below, this would be 25/4 = 6.25 ml, which would be most suitably added from a burette. Add 25 ml of the vanadium molybdate reagent and make up to the mark with water (both liquids at 20°C), mix and stand for 10 min.

Using a matched pair of cells, place the lowest standard in the reference beam, and measure the absorbance of the other standards at 420 nm. Plot the standard curve relating absorbance to known concentration.

Procedure (analysis). Dilute the basic slag extract to give (≥25 ml) a solution containing 220–248 µg P_2O_5 ml^{-1} (96–108 µg P ml^{-1}) at 20°C. Thus, if the basic slag contains 920 µg P_2O_5 ml^{-1}, a dilution of 25 ml extract to 100 ml will yield a solution of 230 µg P_2O_5 ml^{-1} in 0.5% (m/v) citric acid. Pipette 25 ml of this solution into a 100-ml volumetric flask, add 25 ml vanadium molybdate (20°C), make up to the mark with water (20°C), mix and stand for 10 min. Simultaneously prepare a fresh 50 µg P_2O_5 ml^{-1} (21.8 µg P ml^{-1}) reference standard containing the same amount of citric acid as the diluted sample extract, against which the absorbance of the samples is measured (see 'Procedure (standard curve)' above). Maintain all solutions at 20°C.

Calculation. Divide the concentration of the sample solution read from the standard curve by 3 (if diluted as in above sample analysis procedure), to obtain the % P, or divide by 1.304 to obtain the % P_2O_5 in the sample.

Example: if the diluted sample concentration as read from the standard curve was 60 µg P_2O_5 ml^{-1}, this corresponds to a concentration of:

$$60 \times 100/25 \times 100/25 = 960 \text{ µg } P_2O_5 \text{ ml}^{-1} = 960 \times 0.4364 = 418.9 \text{ µg P}$$

in the original extract (because 25 ml of citric acid extract was diluted to 100 ml, and 25 ml of this solution was taken and further diluted to 100 ml). There was 500 ml of original citric acid extract, therefore this volume would contain:

$$500 \times 960 \text{ µg } P_2O_5 = 480,000/10^6 \text{ g } P_2O_5 = 0.480 \text{ g per 5.0 g sample,}$$
or 9.60% P_2O_5 in the original sample.

This is equivalent to 9.60 × 0.4364 = 4.19% P. Since 60 µg P_2O_5 ml^{-1} in the final sample solution corresponds to 9.60% P_2O_5 in the original sample, sample concentrations from the standard curve should be multiplied by 9.60/60 = 0.16 to give the % P_2O_5, or 0.07 to give the % P in the original sample.

Method 6.2c. Determination of total phosphorus in the acid digest from Method 6.1b. with final determination by autoanalysis

The acid digest from Method 6.1b consists of a solution of 0.05 g sample in 50 ml of 50% v/v H_2SO_4. There are therefore 50,000 µg sample in 50 ml, or 1000 µg ml^{-1}. If the fertilizer is a super triplephosphate with approximately 20% P, this would give a solution of approximately 200 µg P ml^{-1}. The auto-analysis method is the one described in Chapter 7 for total phosphorus in plant materials, which requires an optimum sample concentration of from 20 to 100 µg P ml^{-1}, therefore 5 ml of the super triplephosphate sample solution should be diluted with 50% v/v H_2SO_4 to 20 ml in a volumetric flask to give a solution of approximately 50 µg P ml^{-1}. The minimum phosphorus content in the fertilizer to give an adequate peak height is 2% P, or 4.6% P_2O_5. Single superphosphate has approximately 8% P, so the resulting 80 µg P ml^{-1} solution should not need further dilution. As can be seen from the above discussion, the number of µg P ml^{-1} read from the standard curve is divided by 10 to give the % P in the sample of fertilizer.

Discussion 6.3. Determination of potassium in fertilizers

If the fertilizer is a straight potassium fertilizer such as potassium chloride (muriate of potash), sulphate or nitrate, it may be extracted by shaking with water. For any fertilizer, including mixed or compound fertilizers, the AOAC official method (No. 983.02B, Preparation of sample) recommends extraction with boiling 4% ammonium oxalate solution. The ammonium oxalate was introduced to precipitate calcium, which could suppress the potassium reading by up to 2.5% in an erratic fashion (Schwer and Conan, 1960). The precipitated calcium oxalate is allowed to settle and an aliquot for analysis taken from the supernatant, or else it is filtered. The final determination uses an automatic analyser and flame photometer (Johnson, 1990b). The SI 1996 No. 1342, however, recommends extraction of straight, compound or fluid fertilizers with boiling water followed by removal of interferences and gravimetric determination of precipitated potassium tetraphenylborate. We will give both extraction methods, and assume the final analysis is by flame photometry. Results using the AOAC method should be reported as oxalate-potassium, thus differentiating them from water-soluble-potassium, which might not give the same values. The one drawback with using flame photometry in fertilizer analysis is the 'phosphate effect'. This is the depression of the intensity of the potassium flame emission signal arising from the presence of higher amounts of phosphate in the fertilizer (Gehrke *et al.*, 1964). These authors found it was the only commonly encountered ion that could cause a serious interfering effect. However, at the normal phosphate levels found in superphosphate, mono- and diammonium phosphate fertilizers, no interference was found by Schwer and Conan (1960). Gehrke *et al.* (1964) also found that calcium enhanced the signal if present at more than 50 µg Ca ml^{-1} (but see Schwer and Conan, 1960, who found a depressive effect).

If the Ca:K ratio exceeds 10:1, the interference is significant and the Ca should be removed, which can be achieved by extraction and precipitation using ammonium oxalate. The phosphate effect on a 20 µg K ml^{-1} standard was such that there was no effect on the K:Li emission intensity (Li is the internal standard for the Technicon flame photometer used by the authors) for concentrations up to 50 µg P_2O_5 ml^{-1}, an enhancement occurred from 50 to 200 µg P_2O_5 ml^{-1}, no further effect from 250 to 350 µg P_2O_5 ml^{-1}, but a suppression occurred from 350 to 750 µg P_2O_5 ml^{-1}, when a further plateau occurred. This can be largely compensated for by making up the potassium standards with potassium dihydrogen orthophosphate (potassium phosphate, monobasic). The AOAC method, in addition, adds a lanthanum solution (La_2O_3 in HNO_3) to the lithium nitrate internal standard reagent in order to remove the phosphate effect.

In the manufacture of fertilizers, the large numbers of samples prohibit the routine use of the official methods for on-line analysis of potassium. Instead, Hydro Agri (UK) Ltd use a modification of the AOAC method with air-segmented continuous flow analysis and detection by flame photometry using an internal standard. In the UK, official methods in The Fertilisers (Sampling and Analysis) Regulations 1996 – SI 1996 No. 1342 are used to check that the marketed fertilizer conforms to the Fertiliser Regulations 1991 and subsequent amendments (J. Vessey, Hydro Agri (UK) Ltd, 2001, personal communication).

If the acid digestion for total nitrogen has already been carried out according to Methods 6.1a or 6.1b, either of the diluted digest solutions may be used, providing standards are made up in the same concentration of sulphuric acid as the extract. Both methods give the same concentration of potassium in the final diluted solution: a pure potassium chloride sample will yield 524 µg ml^{-1} K, and a pure potassium nitrate sample gives a solution of 387 µg ml^{-1} K. With any of the methods, further dilution may be necessary to suit the sensitivity of the flame photometer, and the calculation should be amended accordingly.

Method 6.3a. Determination of water-soluble potassium

Reagents.

- Potassium stock standard solution, 1000 µg K ml^{-1} – dry potassium dihydrogen orthophosphate (KH_2PO_4) for 2 h at 105°C, then allow to cool in a desiccator. Weigh 2.889 g into a 100-ml beaker, add sufficient water to dissolve, and transfer to a 1-l volumetric flask with beaker washings; make up to the mark with water and mix.
- Potassium working standard solutions, 0–400 µg K ml^{-1} – pipette 0, 5, 10, 20, 30 and 40 ml potassium stock standard solution into a series of 100-ml volumetric flasks and make up to the mark with water and mix. This will provide a series of standards containing 0, 50, 100, 200, 300, and 400 µg K ml^{-1}.

Procedure. Weigh 5.000 g of the powdered sample into a 600-ml beaker, then add approximately 400 ml water, and cover with a watch glass. Place on a hotplate and bring to the boil, then continue boiling gently for 30 min. Allow to cool, then transfer, with washings, to a 1-l volumetric flask, make up to the mark and mix. Filter into a dry sample container, rejecting the first 50 ml filtrate. Pipette a 10-ml aliquot into a 100-ml volumetric flask, make up to the mark and mix. Analyse using a flame photometer, set up according to the manufacturer's instructions. Compare with a series of standards containing 0, 50, 100, 200, 300 and 400 µg K ml^{-1}.

Calculation. As an example, 100 µg K ml^{-1} in the final sample solution would correspond to 1 g K (20% K) in the original 5 g sample. (Because of the final ×10 dilution, 100 µg K ml^{-1} gives $10 \times 100 \times 1000$ µg K in the 1 l of sample solution. Thus there is 1.000 g K in a solution of 5.000 g sample, which amounts to 20% K.) Therefore divide the µg K ml^{-1} by 5 to get the % K in the sample of fertilizer. (Pure potassium chloride will give a final sample solution of 262.2 µg K ml^{-1}.) Multiply the % K by 1.2047 to get the equivalent value of % K_2O.

Method 6.3b. Determination of ammonium oxalate-soluble potassium

Reagents.

- Ammonium oxalate, 4% (m/v)
- Potassium standard solutions – see 6.3a.

Procedure. Weigh 1 g of the powdered sample into a 600-ml beaker, then add 50 ml 4% (m/v) ammonium oxalate solution, 125 ml water and cover with a watch glass. Place on a hotplate and bring to the boil, then boil gently for 30 min and allow to cool. Transfer, with beaker washings, to a 500-ml volumetric flask and make up to the mark with water and mix. Filter into a dry sample container, rejecting the first 50 ml of filtrate. Pipette 25 ml of this solution into a 100-ml volumetric flask, make up to the mark with water and mix. Analyse using a flame photometer, set up according to the manufacturer's instructions. Compare with a series of standards containing 0, 50, 100, 200, 300 and 400 µg K ml^{-1}.

Calculation. A concentration of 100 µg K ml^{-1} in the final sample solution corresponds to 0.2 g K (20% K) in the original 1 g sample. (Because of the final × 4 dilution, 100 µg K ml^{-1} gives $4 \times 100 \times 500$ µg K in the 500 ml of sample solution. Thus there is 0.200 g K in a solution of 1.000 g sample, which amounts to 20% K.) Therefore divide the µg K ml^{-1} by 5 to get the % K in the sample of fertilizer. (Pure potassium chloride will give a final sample solution of 262.2 µg K ml^{-1}.) Multiply the % K by 1.2047 to get the equivalent value of % K_2O.

Method 6.3c. Determination of potassium in the acid digest from Methods 6.1a. or 6.1b.

Reagents.

- Potassium standard solutions – see Method 6.3a.

Procedure. Using a pipette filler, pipette 25 ml of the acid digest solution from Method 6.1a (in 10% v/v H_2SO_4) or Method 6.1b (in 50% v/v H_2SO_4) into a 50-ml volumetric flask, make up to the mark with water and mix. Analyse using a flame photometer, set up according to the manufacturer's instructions. Compare with a series of standards made up either in 5% (v/v) H_2SO_4 or 25% (v/v) H_2SO_4 as appropriate, containing 0, 50, 100, 200, 300, and 400 µg K ml^{-1}.

Calculation. A concentration of 100 µg K ml^{-1} in the final diluted digest solution corresponds to 0.025 g K (10% K) in the original 0.25 g sample for the sample digest solution from Method 6.1a, and corresponds to 0.005 g K (10% K) in the 0.05 g sample for the sample digest solution from Method 6.1b. Therefore, divide the µg K ml^{-1} by 10 to get the % K in the sample of fertilizer. (Pure potassium chloride will give a final sample solution of 262.2 µg K ml^{-1}.) Multiply the % K by 1.2047 to get the equivalent value of % K_2O.

Liming Materials

The Fertilisers Regulations 1990 Group 5(a) describes about 21 types of liming material, and Group 5(b) covers any not specified in Group 5(a). These regulations set limits for the content of MgO and the percentage which will pass through various sieve mesh sizes. They also give the essential value/s that must be declared. This is always the neutralizing value, and sometimes the percentages passing through various sieve sizes. An example is magnesian (US: magnesic) ground limestone. The meaning of this term is defined as 'Sedimentary rock consisting largely of calcium and magnesium carbonates and containing not less than 15% of magnesium as MgO and of which 100% will pass through a sieve of 5 mm, not less than 95% will pass through a sieve of 3.35 mm and not less than 40% will pass through a 150 micron sieve.' The declared values are 'Neutralizing value', and 'Amount of material as a percentage by weight that will pass through a 150 micron sieve.' The latter value is commonly called a determination of fineness of grinding. The stated values may vary by 5% from the measured values. We will describe the method for determining moisture, neutralizing value, and fineness of grinding as given in the SI 1996 No. 1342, The Fertilisers (Sampling and Analysis) Regulations 1996, Schedule 2 Part II, [5.2. Determination of moisture, 5.6. Determination of the neutralizing value in liming materials, 5.7. Determination of fineness of products other than potassic basic slag], and indicate in brackets where the AOAC method differs. The AOAC methods for agricultural liming materials are given by Johnson (1990a), and include methods for assessing carbon dioxide, silica and oxides of aluminium, iron, phosphorus, and titanium; also elemental aluminium, calcium, iron, manganese, magnesium, phosphorus, silicon and sulphur as sulphide.

Method 6.4. Determination of the moisture and neutralizing value of liming materials

Reagents.

- Hydrochloric acid, 0.5 M
- Sodium hydroxide, 0.5 M, carbonate free
- Phenolphthalein indicator – dissolve 0.25 g phenolphthalein in 150 ml 95% v/v ethanol and dilute to 250 ml with water.

Procedure (moisture determination). From about 2 kg of bulk sample, cone and quarter or by other means obtain a representative sample of about 200 g. If determination of fineness is to be carried out, duplicate approximately 100-g samples should be taken and any soft lumps disintegrated by lightly crushing. The one for fineness should not be ground further; the other should be rapidly ground until it completely passes a 1.0 mm sieve (test sieves conforming to British Standard 410: 1986 are suitable). Mix the ground sample well and form into a flattened cone. Taking random portions with a spatula, weigh 5 g of the prepared sample to the nearest 0.001 g, and transfer to a previously weighed container with airtight lid. Place the uncovered container and the lid in an oven and maintain at 100°C (AOAC: 110°C) for 2–3 h. Replace the lid on the container, remove from the oven and allow to cool in a desiccator and weigh. Reheat for another hour, cool and reweigh. If the difference in weight exceeds 0.01 g continue the heating and cooling procedure until a weight constant within 0.01 g is attained. Calculate the total loss of weight and express it as a percentage of the original weight, which gives the percentage moisture in the fertilizer sample as received.

Procedure (neutralizing value). At the same time as weighing the sample for moisture determination, weigh 0.5 g (AOAC: 1 g, but 0.5 g for CaO or $Ca(OH)_2$), or x g, of the prepared sample to the nearest 0.001 g, and transfer to a 300-ml (AOAC: 250-ml) conical flask. Add 50 ml of 0.5 M hydrochloric acid, cover the flask with a watch glass and boil the contents gently for 5 min. Cool the mixture to room temperature, add 2 or 3 drops of the phenolphthalein indicator and titrate with 0.5 M sodium hydroxide solution to the end point of the indicator.

Calculation. Determine the amount of hydrochloric acid (y ml) consumed by the sample. This is done by subtracting the titre of 0.5 M NaOH from 50 (the volume of 0.5 M HCl added to the liming material). The reactions for HCl on limestone and the subsequent back-titration of the excess HCl are:

$$CaCO_3 + 2HCl = CaCl_2 + H_2O + CO_2\uparrow \; ; \; HCl + NaOH = NaCl + H_2O$$

From the reaction equations, 1 mol of $CaCO_3$ (molar mass 100.087) is neutralized by 2 mol of HCl. A volume of 1 l of 1 M HCl will neutralize 100.087/2 g $CaCO_3$ = 50.04 g $CaCO_3$. Therefore, 1 l of 0.5 M HCl will neutralize 25.02 g $CaCO_3$, and 1 ml 0.5 M HCl will neutralize 0.02502 g $CaCO_3$. The result has to be expressed in terms of CaO (molar mass 56.077). Therefore, 1 ml 0.5 M HCl will neutralize the equivalent of $0.02502 \times 56.077/100.087$, or 0.01402 g CaO.

The neutralizing value is expressed as a percentage by weight of calcium oxide (CaO) and refers to the undried sample as received. Thus the formula becomes:

Neutralizing value = $y \times 0.01402/x \times 100\%$ CaO,
where x is the sample weight and y is the titre.

Method 6.5. Determination of fineness of grinding (150 μm/100 mesh fraction)

Thoroughly mix the unground approximately 100 g duplicate sample prepared in Method 6.4. Heat this portion at 100°C until dry and thoroughly mix. Weigh 20 g, to the nearest 0.01 g and transfer to the sieve with the lower receiver attached. Shake the sieve by hand for 5 min, frequently tapping the side. Disintegrate soft lumps such as can be caused to crumble by the application of the fibres of a soft brush, taking care that the hard part of the brush does not make contact with the sieve and that the brush is not used to brush particles through the sieve. Brush out the powder in the lower receiver and weigh. Replace the receiver and repeat the shaking and tapping procedure for 2 min. Add the powder in the receiver to the first portion and weigh. Repeat the process until not more than 0.04 g passes through the sieve during 2 min. Calculate the fineness by expressing the weight of the material passing through the sieve as a percentage of the weight of the portion of the dried sample taken for sieving.

7

The Analysis of Animal Feed and Plant Materials

The official methods applying to Great Britain are found in SI 1999 No. 1663, The Feeding Stuffs (Sampling and Analysis) Regulations 1999. They revoke previous regulations and implement in full the various applicable European Community Directives, and the Community methods of analysis are listed in Schedule 2, part II, Annex I. EC methods in force are usually freely down-loadable as HTML files from the Eur-Lex service, but repealed directives or TIFF versions of current methods from the Official Journal can be searched for and easily purchased by credit card from the Eudor website:

> http://www.eudor.com

These can be sent by FTP (file transfer protocol), e-mail, fax or post. When e-mailed, they will be sent as an attached zipped file. The attached file should be first saved to a directory, then unzipped as a TIFF (tagged image file format) file and a readme.txt file, and re-saved. It is important to use the correct soft-ware to read the TIFF file, which is in tifg4g format. Merely importing into a word-processing package like Microsoft Word 2000 or into a graphics program such as Corel Photopaint™, will only open the first page of the document. By using the Imaging facility in Microsoft Windows Accessories, all the pages may be accessed and printed. The image resolution is selectable when ordering the document, the finest being the default at 400 dpi. The readme.txt file gives details of the helpdesk telephone number and postal address of the Office for Official Publications. Full searches are available to Celex subscribers. TIFF files can also be ordered from any of Eur-OP's document delivery agents listed on:

> http://eur-op.eu.int/general/nl/s-ad.htm

Latest US methods are given by Horwitz (2000), in the 17th edition of *Official Methods of Analysis of AOAC International.* We will refer to the 15th edition (1990) as that is the one at hand, and more likely to be available to others than the latest edition.

Discussion 7.1. Determination of acid detergent fibre, cellulose and lignin

These are a trio of methods, and are described together. The acid detergent fibre (ADF) can be carried out as a separate analysis, but the ADF residue is required to determine the lignin from the weight loss on oxidation by potassium permanganate. Any tannins would also be removed by $KMnO_4$. The residue from the oxidation step contains mainly cellulose (plus any cutin, which appears as dark flecks in the white cellulose) and mineral ash. The cellulose content is determined as the weight loss on ashing. The method is adapted from that of Van Soest and Wine (1968). The AOAC method for lignin uses highly corrosive 72% v/v sulphuric acid instead of permanganate-buffer reagent, and also requires asbestos filter aid, which precludes use of the residue for other determinations. The efficiency of the lignin oxidation depends on the particle size not significantly exceeding the specified mesh. We use 1 mm, but Van Soest and Wine (1968) use 20–30 mesh, approximately 0.5–0.9 mm. It would be worth setting up a spreadsheet for calculations involving large numbers of samples. For further details see Chapter 4, where other aspects of the various fibre determinations are presented more fully.

Method 7.1a. Determination of acid detergent fibre

Reagents.

- Acid detergent solution – add 100 g cetyl trimethylammonium bromide (CTAB), also called hexadecyltrimethylammonium bromide, technical grade, to 2.5 l of 1 M sulphuric acid, stir to dissolve and make up to 5 l
- Sulphuric acid, 1 M – add carefully 56 ml sulphuric acid (approximately 98% m/m H_2SO_4 and 18 M) to about 400 ml water in an 800-ml beaker and stir to dissolve. Transfer with washings to a 1-l volumetric flask, make up to the mark and mix
- Octan-2-ol (also called 2-octanol; capryl alcohol)
- Acetone, commercial 'drum' grade

Procedure. Weigh to the nearest 0.001 g approximately 1 g of oven-dried (⩽65°C) plant material (or air-dried animal feed), ground to 1 mm, into a 500-ml short-neck round-bottomed flask with a ground-glass socket size 34/35. Add 100 ml acid detergent solution, and, if excess foaming is likely to occur, add 1–2 drops octan-2-ol. Place on a macro Kjeldahl heating unit, connect a coil condenser with size 34/35 ground-glass cone, turn on a steady supply of water, and bring to the boil on full heat, then turn the regulator down and allow to simmer for 1 h. Occasionally swirl the flask to wash any sample particles from the flask wall back into the detergent; also, if any bumping has caused sample particles to enter the condenser, squirt a wash-bottle into the top of the condenser to wash them back into the detergent using the minimum amount of water. Filter on a previously dried (100°C) and weighed Porosity 1 sintered glass crucible with gentle suction. The mat of residue is broken up with a small rounded end glass rod and washed twice with water near to boiling point (wear heat resisting gloves). Wash any residue from the sides of the crucible, then wash with successive portions of acetone until no further colour is removed. Suck the residue dry of acetone and allow to stand, preferably in a fume cupboard, until no smell of acetone can be detected, then dry overnight at 100°C. Cool in a desiccator and weigh the crucible plus ADF. Retain the residue for lignin and cellulose determination if required.

Calculation. Subtract the weight of the empty crucible from that of the crucible plus ADF to obtain the weight of ADF. Divide by the sample weight and multiply by 100 to obtain the % ADF in the sample DM (air-dry matter for animal feeds).

Method 7.1b. Determination of lignin

Reagents.

- Potassium permanganate, approximately saturated solution – (*Note*: wear rubber gloves to prevent staining hands.) dissolve 50 g $KMnO_4$ in l l water (solubility = 65 g l^{-1} at 20°C). Store in a brown glass bottle.
- Buffer solution – dissolve 6.0 g ferric nitrate, $Fe(NO_3)_3.9H_2O$, and 0.15 g silver nitrate, $AgNO_3$, in water and make up to 100 ml; add this to a solution of 5.0 g potassium acetate in 500 ml glacial acetic acid in a 2-l beaker and stir to mix. Add 400 ml 2-methylpropan-2-ol (tertiary butyl alcohol, $(CH_3)_3COH$; this solidifies ≤25.5°C, therefore it may need warming to melt before use), and stir to mix. Store in a brown glass bottle.
- Combined permanganate and buffer solution – add two parts by volume saturated $KMnO_4$ solution to one part buffer solution and mix. This will keep for 1 week if refrigerated, therefore only make sufficient for this period's analyses.
- Demineralizing solution – dissolve 50 g oxalic acid $((COOH)_2.2H_2O)$ in 700 ml 95% v/v ethanol. Add 50 ml hydrochloric acid (approximately 36% m/m) and 250 ml water and mix.
- Ethanol, approximately 76% v/v – add 200 ml water to 800 ml 95% v/v ethanol.

Procedure. Wear PPE – rubber gloves, face shield and lab coat when handling the permanganate-buffer solution. Place the weighed sintered glass crucibles containing the ADF residue one at a time into a stainless steel or polythene tray containing 2–3 cm cold water (a 400 × 320 × 50 mm tray will hold 48

crucibles). Holding the crucible, immediately add 25 ml of the combined permanganate and buffer solution to each crucible. Adjust the water level so that
there is only a small diffusion of permanganate-buffer out of the crucibles.
Place a short glass rod (approximately 80 × 4 mm with the ends rounded in
a flame) into each crucible, break up the residue mat and stir to ensure thorough contact of the particles with the reagent; leave the rod in the crucible.
Allow to stand for approximately 90 min at 20–25°C, and top up with
permanganate-buffer as required. The mixture in the crucible should remain
purple; if it turns brown, it is exhausted and should be replaced by fresh
reagent. Filter the crucibles but do not wash at this time. Place the crucibles
in a clean empty tray, half fill with demineralizing solution and stir with the
rod to mix. Allow to stand for 5 min, suck dry, and half refill with demineralizing solution, washing down the sides of the crucible. Allow to stand for
20–30 min until the fibrous residue is white. Filter with gentle suction, then
remove suction and fill with 80% v/v ethanol, stir thoroughly and suck dry.
Repeat this twice more, then wash twice with acetone and suck dry. Allow
to air-dry in a fume cupboard until no smell of acetone is detectable, then
dry overnight in an oven at 100°C, cool in a desiccator, and weigh.

Calculation. Subtract the weight of the crucible plus fibre after oxidation from
the weight of the crucible plus ADF, divide by the initial sample weight and
multiply by 100 to obtain the % lignin in the sample DM (air-dry matter for
animal feeds).

Notes on the reagents. The permanganate reagent oxidizes and dissolves the
lignin and any tannins. The 2-methylpropan-2-ol enhances the wetting of the
ADF fibres by the permanganate. The acetic acid in the buffer solution is to
neutralize the alkali formed in the oxidation reaction:

$$KMnO_4 + 0.5 H_2O = MnO_2 + 1.5 [O] + KOH$$

$$KOH + CH_3COOH = CH_3COOK + H_2O$$

The silver nitrate helps preserve the permanganate solution from decomposition. The oxidation reaction is incomplete below 18°C and too vigorous
above 25°C. The 90 min oxidation stage is not long enough for the complete
removal of lignin from faeces, bark or wood. The demineralizing reagent
removes MnO_2 from the cellulose fibres. Water must be present to remove
the acetic acid and also the MnO_2. Ferric nitrate provides the ferric ion which
prevents precipitation of manganous oxalate, and potassium acetate prevents
free HNO_3 formation from acetolysis of the ferric nitrate.

Method 7.1c. Determination of cellulose and ash

Procedure. Ignite the dried crucible and residue from the lignin determination for 3 h in a furnace at 500°C. Allow to cool in a glass desiccator and
weigh.

Calculations.

1. Percentage cellulose: subtract the weight of crucible and ash from the weight of crucible and residue before ashing to obtain the weight of cellulose. Divide by the sample weight and multiply by 100 to obtain the % cellulose in the sample DM (air-dry matter for animal feeds).
2. Percentage residual ash: subtract the weight of the empty crucible from the weight of crucible plus ash, divide by the sample weight and multiply by 100 to obtain the % residual ash in the sample DM (air-dry matter for animal feeds).

Method 7.2. Determination of crude fibre

The EC official method is described in the Official Journal of the European Communities (EC, 1992), and uses specialized glassware. We will base the method on that of MAFF/ADAS (1986, pp. 90–92), (with Crown Copyright permission), but replace the alcohol and diethyl ether (used for washing the residue) with acetone, which is safer and also used in the EC method. The AOAC method is described by Padmore (1990, pp. 80–82). The traditional method uses 0.313 M NaOH, free from carbonate, but the EC method uses 0.23 M KOH. The method is suitable for plant material or animal feeds, but the following pre-treatments may be necessary, especially for feedstuffs.

Pre-treatments (if required). Samples containing >3% (>5%, EC method) calcium carbonate are pre-treated with excess 0.1 M HCl (3 × 30 ml 0.5 M HCl, EC method). Add the acid to the 3 g (1 g, EC method) weighed sample in a 500-ml short-neck round-bottom flask with ground glass socket size 34/35, and swirl for 1 min. Allow to settle, and then decant the supernatant into a 125-mm Whatman No. 541 filter paper, and wash the residue twice with water, decanting the washings into the filter paper. Allow both the residue and the filter to drain thoroughly. Bring 200 ml of 0.128 M sulphuric acid to boiling point, use a portion to wash any particles from the filter paper back into the flask, then add the remainder of the acid to the flask and proceed with refluxing as detailed below.

Samples containing >10% crude fat must be defatted before analysis with petroleum spirit (light petroleum, 40–60°C boiling range). This may be done either in a Soxhlet extractor, or in a beaker by stirring, settling and decanting three times with 30 ml petroleum spirit. Allow the fat-free sample to air-dry in a fume cupboard.

Reagents.

- Acetone, commercial 'drum' grade
- Hydrochloric acid, approximately 0.1 M – dilute 1 vol. of hydrochloric acid, approximately 36% m/m HCl to 100 vol.
- Petroleum spirit, (light petroleum), boiling range 40–60°C
- Octan-2-ol (*n*-octanol)

- Sodium hydroxide, 0.313 M (1.25% m/v NaOH)
- Sulphuric acid, 0.128 M (1.25% m/v H_2SO_4)

Procedure. Weigh 3 g of either a feed sample (as received, ground to 1 mm; also take a separate sample at the same time for moisture analysis) or plant sample dried at ≤65°C, into a 500-ml short neck round-bottom flask with ground glass socket size 34/35. Add 200 ml of 0.128 M sulphuric acid (measured at room temperature) which has been heated to boiling point, place the flask on a macro Kjeldahl heating unit, connect a coiled reflux condenser through which a steady stream of cold water is flowing, and bring to the boil within 1 min on full heat. Reduce the heat and continue to boil gently for 30 min. Reduce any excessive foaming by addition of 1–2 drops octan-2-ol, and swirl every 5 min to wash sample particles from the flask wall back into the acid. If any bumping has caused sample particles to enter the condenser, squirt a wash-bottle into the top of the condenser to wash them back into the acid using the minimum amount of water. Assemble a borosilicate glass Hartley type three-piece funnel with polypropylene support plate (e.g. Whatman, 530 ml, 125 mm) on which is a 125-mm Whatman No. 541 filter paper. Preheat the funnel by pouring boiling water into it. Turn the heat under the flask off and allow to stand for approximately 1 min before pouring into a shallow layer of hot water in the funnel. Adjust the suction so that the filtration is completed in less than 10 min. Wash the insoluble matter with boiling water until the washings are neutral to litmus paper. Wash the residue back into the flask using 200 ml (measured at room temperature) of boiling 0.313 M sodium hydroxide. Boil for 30 min, as described previously for the acid, then allow to stand for 1 min, and filter hot through a 60-ml, porosity No. 1 (P160, 100–160 μm) sintered glass crucible using gentle suction. Transfer the whole of the insoluble material from the flask to the crucible with hot water. Wash first with boiling water, once with approximately 0.1 M HCl, and then with water until the washings are neutral to litmus paper. Wash with three successive 25 ml amounts of acetone and suck dry, then air-dry in a fume cupboard until no smell of acetone can be detected. Dry the crucible and contents to constant weight in an oven at 130°C (as recommended by EC and AOAC methods), allowing to cool in a desiccator before weighing. Place the crucible in a cool muffle furnace, and increase the temperature to between 475°C and 500°C; keep at this temperature for at least 30 min and until ashing is complete. It is important not to overshoot the temperature, because the sintered crucibles will be damaged at ≥ 515°C. Remove the crucible from the muffle furnace, cool in a desiccator and weigh.

Calculation. Divide the loss in weight on ignition by 0.003 to give the g kg^{-1} crude fibre in the sample, or divide by 0.03 to give the % crude fibre in the sample.

  Note: The EC method also specifies a blank determination, where the weight loss on ashing should not exceed 4 mg. This should be subtracted from the loss in weight of the sample on ignition when calculating the % crude fibre.

Method 7.3. Determination of modified acid detergent fibre (MAD fibre)

This modification to the ADF method was designed to improve the relation-ship between the ADF and digestibility in ruminants. The sample drying tem-perature of 95°C, however, means it is unsuitable for assaying heat damage and unavailable protein (Van Soest, 1982). The method below is based on that given in MAFF/ADAS (1986, pp. 93–94), with Crown Copyright permis-sion. See Chapter 4 for further discussion on fibre extraction procedures.

Reagents.

- Acetone, commercial 'drum' grade
- Octan-2-ol (*n*-octanol)
- Sulphuric acid-CTAB solution – dissolve 10 g of cetyltrimethylammonium bromide in 1 l of 0.5 M sulphuric acid using a magnetic stirrer, and filter if necessary.

Procedure. Transfer 1.000 g of the dried (95°C) and ground (1 mm) sample into a 500-ml short-neck round-bottom flask with ground glass socket size 34/35. Add 100 ml of the sulphuric acid-CTAB solution, transfer to a macro Kjeldahl heating unit and connect a coiled reflux condenser through which a steady stream of cold water is flowing, and bring to the boil within 1 min on full heat. Reduce the heat and continue to boil gently for 2 h. Reduce any excessive foaming while boiling, or at the filtration stage, by addition of 1–2 drops octan-2-ol, and swirl every 5 min to wash sample particles from the flask wall back into the acid. If any bumping has caused sample particles to enter the condenser, squirt a wash-bottle into the top of the condenser to wash them back into the acid using the minimum amount of water. Filter hot through a previously dried (place in cool muffle, raise to 500°C and main-tain for 30 min) and weighed 60 ml borosilicate porosity No. 1 (P160, 100–160 µm) filter crucible using gentle suction. Wash the residue with 3 × 50 ml portions of almost boiling water and then with acetone. Allow to air-dry in a fume cupboard until no smell of acetone can be detected. Dry the crucible and contents overnight in an oven at 102°C. Allow to cool in a des-iccator and weigh. Retain the crucible and contents for determination of ash-free MAD fibre, as detailed below.

Procedure (ash-free MAD fibre). Place the crucible and its contents of MAD fibre in a cool muffle furnace. Increase the temperature of the furnace to 475–500°C; keep at this temperature for at least 30 min and until ashing is complete. It is important not to overshoot the temperature, because the sin-tered crucibles will be damaged at $\geqslant$ 515°C. Remove the crucible from the muffle furnace, cool in a desiccator and weigh.

Calculations. Subtract the weight of the empty crucible from the weight of the crucible plus MAD fibre and multiply by 1000 to give the g kg^{-1}, or by 100 to give the % MAD fibre in the sample DM.

Subtract the weight of the crucible plus ash from the weight of the crucible plus MAD fibre and multiply by 1000 to give the g kg^{-1}, or by 100 to give the % ash-free MAD fibre in the sample DM.

Method 7.4. Determination of neutral cellulase plus gamanase digestibility (NCGD) of feeding stuffs

The method is based on that of MAFF (1993b) and is discussed in Chapter 4. It is intended for samples of compound feeds or feed mixtures.

Apparatus.

- Filter tubes – a special sintered borosilicate glass filter tube with Suba-Seal and plastic cap is available from Soham Scientific, Unit 6, Mereside, Soham, Ely, Cambs CB7 5EE, UK, and is shown in Fig. 7.1.

Reagents.

- Enzymes – test kits of consistent quality are obtainable from Biotal Limited, 5 Chiltern Close, Cardiff, CF4 5DL, UK; tel. +44 (0)2920 766716, fax +44 (0)2920 747414. The cost is approximately £327 + VAT per kit for 250 tests.
- Acetate buffer solution, pH 4.8 – dissolve 1.36 g of sodium acetate in 500 ml of distilled water, add 0.6 ml glacial acetic acid and dilute to 1 l. Check the pH now and before use, and adjust to pH 4.8 with sodium hydroxide solution.
- Acetone, commercial 'drum' grade.
- Amylase solution – dissolve 2 g of α-amylase in 90 ml distilled water and filter. Add 10 ml of 2-ethoxyethanol to the filtrate, and store at 5°C. Prepare fresh daily.

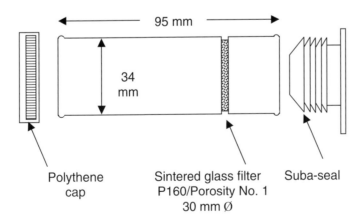

95 mm

34 mm

Polythene cap

Sintered glass filter P160/Porosity No. 1 30 mm Ø

Suba-seal

Fig. 7.1. Borosilicate tube with sintered glass filter and end seals.

- Buffered cellulase solution – transfer 20 g cellulase (also called polysaccharase in the kit) and 0.1 g chloramphenicol to a 2-l wide-neck conical flask. Add 1 l of acetate buffer solution, shake and incubate for at least 1 h at 40°C.
- Cellulase-gamanase solution – add nine volumes of cellulase solution to one volume of the gamanase preparation and mix thoroughly. Filter through a Whatman GF/A glass micro-fibre filter circle held in a Hartley type three-piece funnel.
- Chloramphenicol, (D(-)-threo-2-dichloroacetamido-1-*p*-nitrophenylpropane-1,3-diol; $C_{11}H_{12}Cl_2N_2O_5$, molar mass 323.13). *Note*: ingestion may cause a toxic effect on bone marrow, and contact with skin may cause an allergic response.
- Neutral detergent solution – dissolve 93 g EDTA disodium salt, and 34 g disodium tetraborate decahydrate ($Na_2B_4O_7.10H_2O$) in water on a stirrer/hotplate. Add 150 g sodium dodecyl sulphate (sodium lauryl sulphate, $CH_3(CH_2)_{11}OSO_3Na$)) and 50 ml of triethylene glycol ($HO(CH_2)_2O(CH_2)_2O(CH_2)_2OH$). (*Note*: the published method specifies the toxic 2-ethoxyethanol ($CH_3CH_2OCH_2CH_2OH$)). Add a solution of 22.8 g of sodium dihydrogen phosphate, anhydrous (NaH_2PO_4) prepared by dissolving in water on a stirrer/hotplate. Dilute to 5 l and mix thoroughly. Adjust the pH to 6.9–7.1 if necessary. *Note*: sodium dodecyl sulphate dust irritates the lungs, therefore wear dust masks when weighing and use dust extraction fans.
- Petroleum spirit (light petroleum), boiling range 40–60°C

Procedure. Prepare a filter tube by placing in a cool muffle furnace, increase the temperature to 500°C, and maintain it for 30 min. Remove and allow to cool in a desiccator. To avoid damaging the sinter, do not allow the temperature to reach 515°C. Weigh 0.5 g of the sample, as received, but ground to 1 mm, into the crucible. At the same time, weigh a separate portion for DM and ash determination. Wash the sample in the filter tube with 3 × 25 ml portions of petroleum spirit. Suck as dry as possible under gentle vacuum and complete the removal within 10 min at 60°C in an oven specified for use with flammable solvents.

Carefully brush the fat-free residue into a 150-ml flat-bottom flask, add 25 ml of neutral detergent and swirl to mix. Place in the heating unit and attach the reflux condenser, ensuring a steady flow of cold water. Bring to the boil and reflux for 30 min. Swirl occasionally to prevent overheating the sample and ensure adequate mixing. Then turn off the heat and add 25 ml of cold neutral detergent solution followed by 2 ml of amylase solution. Again heat to boiling and reflux for a further 30 min, occasionally swirling the contents of the flask.

Turn off the heat and immediately filter through the same filter tube previously used for the fat extraction. Wash the residue thoroughly with at least 3 × 20 ml hot distilled water, as it is essential to remove all the neutral detergent solution. Moisten a Suba-Seal and carefully push it into the bottom end of the filter tube as far as the sinter disc. Use dispensers to add 25 ml of distilled water at 80°C followed by 2 ml of amylase solution. The force of the jet from the dispensers should be enough to agitate the contents; if not, perform an additional mixing; allow to stand for 15 min. Remove the Suba-Seal and apply suction to remove the amylase solution. Replace the Suba-Seal and, using a dispenser, add 30 ml buffered cellulase-gamanase solution to the residue. Secure the polythene cap on the top of the tube and shake to mix the fibres thoroughly with the enzyme solution. Incubate at 40°C (± 2°C) for 40 h, shaking morning and evening to ensure adequate mixing.

Remove both the Suba-Seal and polythene cap and place the filter tube in the rubber cone adapter (42 × 27 mm) used with the adapter funnel attached to the Buchner flask, or other suitable device. Wash any particles from the polythene cap into the filter tube, apply just sufficient suction to remove the gamanase-cellulase solution, then wash the residual undigested fibre with hot distilled water (approximately 80°C). Finally wash well with acetone, leave to air dry in a fume cupboard, and when no smell of acetone can be detected, dry in an oven overnight at 100°C (±2°C). Cool in a desiccator and weigh the filter tube plus residue.

Place the crucible and its contents of NCGD fibre in a cool muffle furnace. Increase the temperature of the furnace to 500 ±5°C; keep at this temperature (for approximately 3 h) until ashing is complete. It is important not to overshoot the temperature, because the sintered disc will be damaged at ≥515°C. Remove the crucible from the muffle furnace, cool in a desiccator and weigh.

Calculation. Subtract the weight of crucible plus ash from the weight of crucible plus residue to obtain the weight of indigestible organic matter in the 0.5 g sample ('as received'). Divide by the sample weight and multiply by 100 to get the % indigestible organic matter in the 0.5 g sample ('as received'). This must be corrected for moisture in the sample, therefore using the value from a separate moisture determination, multiply by 100/(100–%moisture) to obtain the % indigestible organic matter in sample DM. From the total ash determination, calculate the % total ash, and correcting for sample moisture as above, express the total ash as % in sample DM. The NCGD = 100 – (% indigestible organic matter in DM + % total ash in DM).

Method 7.5. Determination of neutral detergent fibre (NDF) or plant cell-wall constituents

This is based on the method by Van Soest and Wine (1967) which has been modified according to subsequent recommendations. It is the only fibre determination suitable for non-ruminants. The residue consists of the plant cell-wall constituents: cellulose, hemicellulose, lignin, cutin, NDF-insoluble tannin and ash. See the article by Cherney (2000) for current modifications; these include the use of amylase to aid in the removal of starch from forages containing grain (Van Soest *et al.*, 1991), which has been adopted by MAFF

(1993c). Also, the use of sodium sulphite and decalin has been eliminated; see Chapter 4 for further discussion.

Reagents.

- Neutral detergent solution – add 30 g of sodium dodecyl sulphate (sodium lauryl sulphate, $CH_3(CH_2)_{11}OSO_3Na$), 18.61 g of EDTA disodium salt, 6.81 g of sodium borate decahydrate, and 4.56 g of disodium hydrogen phosphate (anhydrous), to 1 l water, and stir to dissolve. Adjust the pH to 6.9–7.1 if necessary. *Note*: sodium dodecyl sulphate dust irritates the lungs, therefore wear dust masks when weighing and use dust extraction fans. If foaming is a problem, also add 10 ml triethylene glycol ($HO(CH_2)_2O(CH_2)_2O(CH_2)_2OH$); the 2-ethoxyethanol previously used is toxic and should be avoided.
- Acetone, commercial 'drum' grade.

Procedure. Weigh 0.5 g of the freeze-dried (or air-dried, but not oven dried) sample, ground to 1 mm, and transfer to a 500-ml short-neck round-bottom flask with ground glass socket size 34/35. (*Note*: if the sample is an animal feed with ⩾10% oil content, it should be defatted with petroleum spirit (40–60°C) before proceeding.) At the same time, weigh a sample for dry matter determination. Add 100 ml neutral detergent solution (ambient temperature), place on a macro Kjeldahl heating unit and connect a coil condenser with size 34/35 ground-glass cone, turn on a steady supply of water, and bring to the boil on full heat, then turn the regulator down and allow to simmer for 1 h. Occasionally swirl the flask to wash any sample particles from the flask wall back into the detergent; also, if any bumping has caused sample particles to enter the condenser, squirt a wash-bottle into the top of the condenser to wash them back into the detergent using the minimum amount of water. Filter on a previously dried (place in cool muffle, raise to 500°C and maintain for 30 min) and weighed Porosity 1 sintered glass crucible with gentle suction (excessive suction compresses the mat and hinders the efficiency of the washing/filtration). Rinse the sample particles from the flask with the minimum amount of very hot water (80–90°C). Remove the vacuum, break up the mat of sample fibres with a small glass rod (approximately 4 mm diam. with ends rounded in a flame) and fill the crucible with very hot water. Apply just sufficient suction to filter, and repeat the washing process. Wash twice with acetone using the same technique and suck dry. Allow to air-dry in a fume cupboard until no smell of acetone is detectable, then dry overnight in an oven at 100°C, cool in a desiccator, and weigh.

Place the crucible and its contents of NDF in a cool muffle furnace. Increase the temperature of the furnace to 500 ±5°C; keep at this temperature (for approximately 3 h) until ashing is complete. It is important not to overshoot the temperature, because the sintered disc will be damaged at ⩾515°C. Remove the crucible from the muffle furnace, cool in a desiccator and weigh.

Calculation. Subtract the weight of the empty crucible from that of the crucible plus NDF to obtain the weight of NDF in 0.5 g sample. Divide by the sample weight and multiply by 100 to obtain the % NDF in the freeze-dried sample. Multiply this figure by 100/(100–moisture content) to obtain the % NDF in DM. Subtract the empty crucible weight from the weight of crucible plus ash and multiply by 100/weight of NDF to obtain the % ash in the NDF.

Method 7.6. Determination of nitrate in plant material by autoanalysis

See the discussion with references in Chapter 4 'Nitrate and water soluble carbohydrate'. Soil extracts low in colour may also be analysed by this procedure by taking a 10-ml scoop of fresh or thawed soil. The official Bran+Luebbe AutoAnalyzer method for nitrate and nitrite in soil, plant and fertilizer extracts is reproduced with permission in Appendix 5.

Reagents.

- Ammonium chloride buffer, pH 7.5 – dissolve 400 g ammonium chloride (NH_4Cl), 40 g EDTA disodium salt, 40 g sodium dihydrogen phosphate dihydrate ($NaH_2PO_4.2H_2O$) and 0.08 g copper sulphate pentahydrate ($CuSO_4.5H_2O$) in 1400 ml previously heated water (5 min in a domestic microwave) contained in a 3-l beaker. Adjust the pH to 7.5 ±0.1 with 10% w/v NaOH and make up to 2 l.
- Benzoic acid solution, saturated – see Chapter 4 'Nitrate and water soluble carbohydrate', for details.
- Cadmium filings – using a wood rasp, file cadmium rod held in a vice, carefully collecting the filings in a surrounding sheet of polythene. Sieve the filings, retaining those between 16 and 25 mesh, and save in a stoppered sample tube. *Note*: cadmium is toxic, so wear gloves and clean the work area carefully afterwards. The filings are used to fill a 380 × 2 mm ID glass tube, plugged at each end for approximately 15 mm with a bunch of fibreglass fibres.
- Orange reagent – warm 1640 ml water containing 200 ml acetic acid, glacial, to approximately 50°C on a magnetic stirrer-hotplate. Add 1.0 g sulphanilamide and stir until dissolved. Add 1.0 g 1-naphthylamine-7-sulphonic acid (Cleve's acid) previously finely ground in a pestle and mortar. Stir to dissolve, using a thick glass rod to crush any remaining particles. Add 280 ml 10% m/v NaOH, cool and adjust the pH to 4.00 ±0.05 using acetic acid or 10% NaOH added dropwise from a disposable polythene Pasteur pipette. Make up to 2 l and store in a dark glass reagent bottle.
- Sodium acetate buffer – dissolve 20 g NaOH in water and add to a solution of 200 ml acetic acid, glacial, in 1400 ml water. Make up to 2 l and adjust the pH to 4.00 ±0.05.
- Stock standard nitrate solution, 100 µg ml^{-1} of NO_3-N – dissolve 0.3034 g sodium nitrate ($NaNO_3$, previously dried) in saturated benzoic acid solution and make up to 500 ml and mix.

• Working standard nitrate solutions, 2–15 µg ml^{-1} of NO$_3$-N – pipette 2, 4, 6, 8, 10 and 15 ml of the stock standard nitrate solution into a series of 100-ml volumetric flasks and make up to 100 ml with saturated benzoic acid solution and mix.

Procedure. Weigh 0.1 g dried herbage into a 250-ml wide-mouth high-density polyethylene screw-cap bottle. The square type bottles fit best the square box of the reciprocating shaker. Add 50 ml saturated benzoic acid solution and shake for 30 min. Filter through a Whatman No. 4 filter paper, rejecting the first few millilitres and save in polythene capped sample tubes. The flow diagram is given in Fig. 7.2.

Switch on the modules for autoanalysis and commence pumping reagents. The wash solution is saturated benzoic acid. Load the standards into a sample tray and analyse at a rate of 20 samples per hour, adjusting the sensitivity of the detection-readout to bring the baseline and top standards on scale. If the samples are all low, the sensitivity should be increased and appropriate lower standards used. After a baseline, analyse a tray of samples plus standards, and follow with a baseline before analysing the second tray of samples and standards. Conclude by aspirating the wash solution to obtain a final baseline.

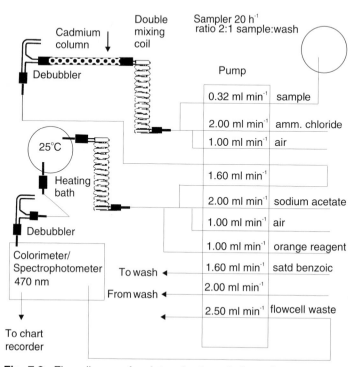

Fig. 7.2. Flow diagram for determination of nitrate in plant materials.

Calculation. Draw a baseline on the chart under all the sample peaks by connecting the baseline from aspirating wash at the start, between trays and at the end. Read the concentration of the sample solutions by comparing the peak heights of the samples with the standards using a chart reader (see Chapter 1, 'Chart reader'). Divide the concentration of the sample solution in µg ml^{-1} of NO_3-N by 20 to obtain the percentage NO_3-N in the sample. If a 10 ml soil sample was extracted into 50 ml saturated benzoic acid solution, the concentration of NO_3-N in the extract should be multiplied by 5 to give the concentration of NO_3-N in µg ml^{-1} of NO_3-N in the fresh soil. Otherwise, divide by 2000 to obtain the percentage NO_3-N in the fresh soil.

Discussion 7.7. Determination of total nitrogen (crude protein) in plant material and feeding stuffs

If it is required to perform the determination by digestion and distillation, refer to Method 5.6b. 'Determination of organic plus ammonium-N by digestion and distillation', but use 2 g oven-dry plant sample ground to 1 mm. Also, for the calculation, multiply the sample titre minus blank titre by 0.35 to give the % N in the sample. Multiply the % N by 6.25 to get the % crude protein. This assumes there are 160 g N kg^{-1} plant protein. Traditional factors for other products are: almonds, 5.18; brazil nuts and peanuts, 5.46; coconuts and tree nuts, 5.30; dairy products, 6.38; wheat, 5.7. *Note*: about 20% of any nitrate present will be included.

The official EC method for crude protein in feeding stuffs may be downloaded from:

http://europa.eu.int/eur-lex/en/lif/dat/1993/en_393L0028.html

This is document 393L0028 (which is presented in the Official Journal No. L179, 22.7.93, p.8), and is the Commission Directive 93/28/EEC of 4 June 1993. The determination of crude protein appears as the Annex and amends Point 2 of Annex I to Directive 72/199/EEC.

The AOAC semi-automated Kjeldahl method for crude protein in animal feed with determination by segmented flow autoanalysis is given by Padmore (1990, pp. 72–74). The method below is a modified version of that in Faithfull (1971a), which allowed several elements to be determined simultaneously. Two sample probes may be placed in the sampler dipper, and the capillary from each may be split to supply two different chemistries (auto-analyser modules) per pump. Although this is an efficient system, there are two drawbacks: a fault with the plumbing of one chemistry means the other one going through the same pump will have to be stopped in order to effect a repair; and if there are too many pump tubes attached to the pump, there is a danger that the fluid will stop flowing in one of the tubes because of the increased pressure required between the rollers and the pressure-plate to ensure that the bores of all the tubes are properly compressed. We will describe the single chemistry system, but point out that the addition of another chemistry utilizing the same digest solution is possible.

Method 7.7a. Determination of total nitrogen (crude protein) in plant material by autoanalysis

Apparatus.

- Acid digestion unit – see Chapter 4 for details of aluminium blocks and hotplate.
- Autoanalysis modules for segmented flow system – sampler, pump, manifold, colorimeter (with 640 nm interference filters) or spectrophotometer with 10-mm flowcell, and chart-recorder or computer readout.
- Dispenser, 5 ml.
- Digestion test tubes – 150 × 16 mm diameter, heavy wall (BS 3218) borosilicate glass rimless type; (Fisher Cat. No. TES-674-150S). These should have graduation lines inscribed at 5 ml and 10 ml positions.
- Tongs, stainless steel.

Reagents.

- Acid-digest solution, 0.4% selenium in sulphuric acid (approximately 98% m/m H_2SO_4) – wear PPE. See Chapter 4 for full details of making this highly corrosive solution, and note the safety precautions.
- Acid-wash solution – wear PPE. Add 250 ml sulphuric acid (approximately 98% m/m H_2SO_4) to 250 ml water slowly with stirring. Allow to cool, then make up to 500 ml with water and carefully invert to mix.
- Sodium hypochlorite solution – dilute 800 ml of sodium hypochlorite solution (Merck, 12% w/v available chlorine) to 2 l with water.
- Sodium phenate solution – dissolve 259 g of sodium hydroxide in approximately 1 l of water and allow to cool. Add 315 ml of 80% m/m phenol solution (Fisher) slowly with stirring. Cool and make up to 2 l; store in a refrigerator.
- Stock standard solution, 2000 µg N ml⁻¹, 200 µg P ml⁻¹, 1600 µg K ml⁻¹, (400 µg Ca ml⁻¹) – omit the Ca if it is unlikely to be required, so as to avoid the precipitation of calcium sulphate in the diluted standards. This combined standard solution can be used for the autoanalysis of P and K, and also provides a similar matrix to the sample digests. Each reagent should be dried at 102°C for 1 h and cooled in a desiccator before weighing. Dissolve 1.3745 g potassium chloride, 0.4393 g potassium dihydrogen phosphate, 4.7162 g ammonium sulphate, (and 0.5000 g calcium carbonate), in sulphuric acid (approximately 98% m/m H_2SO_4) and make up to 500 ml with sulphuric acid.
- Working standards – using a pipette filler, and allowing the pipette to drain thoroughly, measure 5, 10, 15, 20 and 25 ml stock standard into a series of 100-ml beakers, followed by 45, 40, 35, 30 and 25 ml sulphuric acid (approximately 98% m/m H_2SO_4) respectively. Carefully transfer, with rinsing, to a series of 150-ml beakers containing approximately 40 ml water, stir slowly to mix, and allow to cool. Transfer, with rinsing, to a series of 100-ml volumetric flasks numbered 1–5, make up to the mark with water, firmly stopper, and invert carefully to mix. The standards will contain as

follows: No. 1, 100 µg N ml^{-1}, 10 µg P ml^{-1}, 80 µg K ml^{-1}, (and 20 µg Ca ml^{-1}); No. 2, 200 µg N ml^{-1}, 20 µg P ml^{-1}, 160 µg K ml^{-1}, (and 40 µg Ca ml^{-1}); No. 3, 300 µg N ml^{-1}, 30 µg P ml^{-1}, 240 µg K ml^{-1}, (and 60 µg Ca ml^{-1}); No. 4, 400 µg N ml^{-1}, 40 µg P ml^{-1}, 320 µg K ml^{-1}, (and 80 µg Ca ml^{-1}); No. 5, 500 µg N ml^{-1}, 50 µg P ml^{-1}, 400 µg K ml^{-1}, (and 100 µg Ca ml^{-1}). Higher standards up to 1000 µg N ml^{-1}, etc., can be similarly made if required.

Procedure (digestion). For full details see Chapter 4, 'Acid digestion procedure'; a summary is given here. Weigh exactly 0.1000 g oven-dry plant material, ground to 1 mm, into the digestion tube. If the sample is freeze-dried, immediately weigh an extra portion for dry matter determination. A stock sample kept as a control could also be weighed. *Note:* wear PPE for the following stages. Carefully dispense 5 ml of the acid-digest solution into the tube, avoiding too rapid a stream of acid, which may cause loss of sample or spillage of acid. Switch on the fume cupboard extractor–scrubber unit. Using stainless steel tongs, load the samples in batches of about ten into the heating block, which should be within 20°C of the ultimate digestion temperature of 310°C. Remove any tubes in danger of frothing over, and replace after approximately 10 min when frothing has subsided. After about an hour, clean the inside wall of the tubes with a length of 4-mm glass rod, reintroducing any particles back into the acid.

After the digestion period of 4.25 h, remove the tubes to the stainless steel racks and allow to cool in the fume cupboard. Make up to the 5-ml mark with sulphuric acid (approximately 98% m/m H$_2$SO$_4$) using a dropping bottle. Next, carefully add water to the 10-ml mark, forming two layers. Starting at the junction of the layers, slowly oscillate a 4-mm glass rod, flattened into a disc at one end, until the two layers are thoroughly mixed; allow the rod to drain completely by touching the wall of the tube. Do this for all the tubes in the rack, then partially immerse in cold water to cool the tubes. Again make up to the 10-ml mark with water and mix with the rod. If there is to be a delay in analysing the solutions, they should be stoppered to prevent absorption of atmospheric moisture. If calcium is to be determined in the diluted solutions, and it is present at $\geqslant$0.4% Ca in DM, they should be analysed immediately after dilution because CaSO$_4$ will precipitate. It will not precipitate from the undiluted solution.

Procedure (autoanalysis). The flow diagram is shown in Fig. 7.3.
Note 1: the mixing of the sample solution in 50% sulphuric acid with the highly alkaline phenate may cause turbulence, which breaks up the segmenting bubbles, resulting in an irregular bubble pattern. To avoid this, connect a right-angle bend immediately after the horizontal A1 connector, insert a 70-mm length of straight glass transmission tubing and allow it to slope downwards at 45°. Add a right-angle bend followed by a 100-mm length of glass transmission tubing and slope this upwards at about 25°, then insert a U-bend to meet the connection to the double mixing coil. The bends can be varied to suit the arrangement on the platter, and the slope of the straight

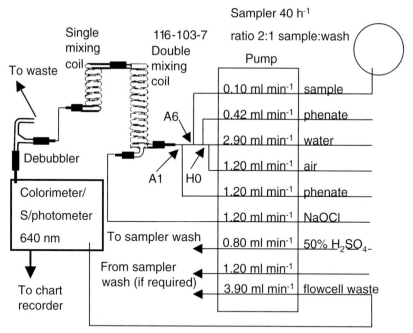

Fig. 7.3. Flow diagram for determination of nitrogen in plant digest solutions in 50% H_2SO_4.

tubes, which affect the mixing dynamics of the reagent solutions of differing densities and viscosities, fine tuned to give the optimum bubble pattern.

Note 2: the stainless steel sample probe is liable to corrode in 50% sulphuric acid. This probe should be raised in the sampler probe holder and merely used to support the polythene capillary, which is attached to it by means of a couple of rings sliced from the end of some PVC transmission tubing (ID 1.6 mm, ⅟₁₆ in). The lower ring should be positioned so that it is just above the liquid level in the wash and sample cups, which prevents cross-contamination. If two polythene capillaries are attached to it, the tips should be held apart and one raised a few millimetres above the other so that a large droplet is not held between them; this would prevent the formation of a small bubble separating sample from wash, and result in a loss of valleys between the peaks.

Switch on the autoanalyser modules and allow pumping of reagents for 30 min to flex the tubes. Carefully pour the sample digest solution from the digestion tube into the sample cup. Use 8.5-ml industrial polystyrene cups or similar size if more than one determination is required on the same sample solution. Load the first sampler tray in the sequence of low to high standards. It is recommended that the lowest standard is in duplicate, and the first peak rejected. After the highest standard, aspirate acid-wash solution for about 5 min to ensure the baseline is reached. Load further trays with about 32 sample cups followed by five standards. The remaining three spare cup positions can be used for higher standards or repeats. Again aspirate acid-wash solution to obtain a baseline between trays; this will enable a correction for any baseline drift to be made. If possible, keep samples of similar analyte concentrations together to avoid interference between adjacent low and high peaks. The sampling rate is set at 40 h^{-1} with a sample:wash ratio of 2:1. When using a chart-recorder, record the sample number on every tenth peak and label the standard peaks; this makes reading the charts easier and enables the identification of problem peaks so that a repeat can be inserted.

Calculation. Draw a baseline under all the peaks by connecting the baselines obtained at the start, between the tray changes, and at the end. Draw a standard curve using a chart reader (see Chapter 1, 'Chart reader'). If exactly 0.1000 g sample was taken, then divide the concentration corresponding to the sample peak, in µg N ml^{-1}, by 100 to give the % total N in DM. Multiply by 6.25 to obtain the % crude protein in DM (but see the exceptions in Discussion 7.7).

If *y* g sample was taken, then (µg N ml^{-1}) × 0.1/100*y* = % N in DM.

Discussion 7.8. Determination of oil in feeding stuffs by extraction with petroleum spirit

The official EC method may be downloaded from:

http://europa.eu.int/eur-lex/en/lif/dat/1998/en_398L0064.html

This is document 398L0064 (which is presented in the Official Journal No. L257, 19.09.98, pp.14–28), and is the Commission Directive 98/64/EC of 3 September 1998. The determination of crude oils and fats appears as Part B of the Annex. Two procedures are described: Procedure A is the directly extractable crude oils and fats applicable to feed materials of plant origin and is a simple extraction with petroleum spirit (light petroleum, boiling range 40–60°C); Procedure B is total crude oils and fats, and applies to feed materials of animal origin and to all compound feeds. It consists of a preliminary heating with 3 M HCl followed by filtration, washing, drying and then Procedure A is carried out. If oil is seen at the filtration stage, then the EC method recommends that Procedure A is carried out first, and then Procedure B. This double extraction procedure, based on SI 1985 No. 1119, is described in MAFF (1993a). Padmore (1990, p. 79) describes the AOAC method for crude fat (or ether extract) in plants or animal feed, which omits the HCl digestion step. We will first describe a basic extraction procedure, then a method suitable for rapeseed. See the discussion in Chapter 4 'Oils, fats and waxes'.

Method 7.8a. Determination of oil in feeding stuffs by extraction with petroleum spirit

Reagents.

- Petroleum spirit, boiling range 40–60°C

Procedure. The sample should not be oven dried, but may be dried over concentrated sulphuric acid or dried at 95–100°C at ≤100 mmHg pressure. A separate sample may be taken for moisture determination. Weigh accurately to ±0.001 g approximately 3–5 g of the sample, ground to 1 mm, into a double thickness extraction thimble, size 22 × 80 mm, and plug the thimble with a wad of oil-free cotton wool. Place a dried (102°C for 1 h) and weighed 250-ml flat-bottom short-neck flask with ground glass 34/35 socket into a recess on the Soxhlet heating unit, and attach the 60 ml size Soxhlet extractor with size 34/35 cone and socket. Place the thimble plus sample into the extractor and pour in sufficient petroleum spirit (40–60°C) so that it just siphons into the flask; then half fill with more solvent and attach the Graham coil reflux condenser with cone size 34/35 (if it is to be left overnight, plug the condenser vents with oil-free cotton wool). Ensure a steady flow of cold water from the condenser, turn on the heater and adjust the control to give a reflux rate of approximately 10 changes per hour; continue refluxing for at least 6 h. Turn the heater off just before the extractor is full, and when it has siphoned out, remove the thimble. Replace the extractor and resume distillation until about 10 ml petroleum spirit-oil solution remains in the flask. The solvent may need to be emptied a couple of times from the reservoir into a waste bottle for redistillation. Turn off the heater, remove the flask and place on a boiling water bath in a fume cupboard to remove the remaining solvent. When no smell of solvent can be detected, place the flask in an oven at 102°C for 2 h, cool and weigh.

Calculation. Subtract the weight of the empty flask from the weight of the flask plus residue of oil, multiply by 1000 and divide by the sample weight to obtain the g kg^{-1} of oil in the sample. Multiply by 100 instead of 1000 to obtain the percentage oil. If required, use the moisture determination to correct to g kg^{-1} oil in DM. Explanation: weight of oil divided by sample weight is the number of grams of oil per gram of sample. Therefore, multiplying by 1000 gives the g kg^{-1} oil in the sample.

Method 7.8b. Determination of oil in rapeseed

The species we have dealt with is winter rape (*Brassica napus* L.). See comments in Chapter 4, 'Oils, fats and waxes', and the reference to Hughes (1969).

Oilseeds with hard shells must first be crushed to allow the solvent to reach the seed contents. The hard round rapeseed is easily fractured using a pestle and mortar, but the seed must not be crushed finely, which would spread the oil over the surfaces of the grinder. (The flatter, shiny and softer linseed does not split open cleanly in the same way, and another method is required. We homogenize 4 g seed for 2 min at 11,500–13,000 rpm in a shortened polystyrene sample tube, occasionally tapping the tube and adjusting its height to ensure all the seeds are broken; see also Chapter 2, 'Homogenization'.) The weight of the seed before and after extraction gives the weight of oil extracted by difference, and this procedure permits multiple extractions in one large Soxhlet extractor.

Procedure. Weigh accurately about 5 g of the split rapeseed into a previously dried and weighed 180-mm filter paper circle, which is then neatly and tightly folded and stapled to form a sachet. The weight of the staple/s must be added using the average weight from a number of staples. Tightly pack the sachets on end, and in layers, into a Quickfit EX5/75 600 ml flanged Soxhlet extractor above a 2-l flask positioned on a heating mantle. The extraction assembly must be firmly clamped to a support stand. Fill the reservoir with petroleum spirit (40–60°C) until it just siphons over, then half fill again. Attach a CX5/25 double-surface condenser and ensure a steady flow of coolant water emerges from the condenser. Turn on the heater control, which should be solid-state for safety reasons, and reflux for 16 h. After this period, turn off the heat just before the level in the reservoir has reached the siphon tube. After the solvent has siphoned over, unclamp the flange, remove all the sachets and allow them to drain off. Place them on a large stainless steel tray on a warm hotplate (40–50°C) in a fume cupboard, and allow to dry until no smell of solvent can be detected. Finally, dry in an oven for 1 h at 100°C, cool in a desiccator and weigh immediately. *Note*: filter paper rapidly absorbs atmospheric moisture.

Calculation. Divide the weight lost by the weight of sample and multiply by 100 to obtain the percentage oil in the seed sample. A separate sample for moisture content must be weighed at the same time as the sample for oil extraction if the percentage oil in seed DM is required.

Method 7.9. Determination of pepsin–cellulase digestibility of plant material

See the discussion in Chapter 4, 'Cellulase digestibility'. This method is based on that of Jones and Hayward (1973 and 1975).

Reagents.

- Cellulase solution – dissolve 6.25 g cellulase from *Trichoderma viride* (Merck Ltd) in 1 l citrate-phosphate buffer, pH 4.6, immediately before use.
- Citrate-phosphate buffer, pH 4.6 – mix together 532.5 ml of 0.1 M citric acid and 467.5 ml of 0.2 M Na_2HPO_4.

- Citric acid solution, 0.1 M – dissolve 19.212 g citric acid ($C_6H_8O_7$) in water, and make up to 1 l.
- Disodium hydrogen phosphate solution, 0.2 M – dissolve 28.392 g disodium hydrogen phosphate (Na_2HPO_4) in water and make up to 1 l.
- Pepsin solution – dissolve 2.0 g pepsin 1:10,000 in 1 l of 0.1 M HCl.

Procedure. Weigh 0.200 g of the dried and ground (to 0.75–1 mm) sample into a screw cap McCartney bottle, graduated at 20 ml. Dispense 20 ml pepsin solution, screw on the cap and shake to mix. Incubate at 40°C for 24 h and shake morning and evening. This is facilitated by holding batches of bottles in a galvanized wire-mesh box (made in-house), and simply inverting a few times. Suck out the pepsin using a sintered glass filter stick, porosity 2 (P100). Rinse the stick with a jet of cellulase solution from a wash bottle, and make up to the 20 ml mark and mix. Incubate for a further 48 h at 40°C, shaking by inversion twice daily. Filter the indigestible residue through a 30-ml sintered glass filter crucible, porosity No. 1 (P160), previously oven-dried and weighed. Wash well with water, then with acetone and leave to air-dry. When no smell of acetone can be detected, dry overnight in an oven at 105°C, then cool in a desiccator and weigh.

Calculation. Subtract the weight of the empty crucible from the weight of the crucible plus residue to obtain the weight of undigested sample *W*. The amount solubilized is 0.2 – *W*, therefore the percentage dry matter solubilized (x) = [(0.2 – *W*)/0.2] × 100. This may be correlated with the *in vivo* dry matter digestibility (y) of grasses using the relationship: $y = 0.54x + 35.0$.

Discussion 7.10. Determination of total phosphorus in plant material and feeding stuffs

The official EC method may be downloaded from:
http://europa.eu.int/eur-lex/en/lif/dat/1971/en_371L0393.html
This is document 371L0393 (which is presented in the Official Journal No. L279, 20.12.71, p.7), and is the Second Commission Directive 71/393/EEC of 18 November 1971. The determination of total phosphorus appears as Part III of the Annex. It is a molybdovanadate colorimetric procedure with the absorbance measured at 430 nm.

The AOAC alkalimetric ammonium molybdophosphate and photometric molybdovanadate methods for animal feed are described by Padmore (1990, pp. 87–88), and for plants by Isaac (1990, p. 56). A spectrophotometric molybdovanadate procedure is also described in MAFF/ADAS (1986, pp. 181–182). The official Bran+Luebbe AutoAnalyzer method for phosphate in soil, plant and fertilizer extracts is reproduced with permission in Appendix 6.

We will describe the autoanalysis method based on Faithfull (1971a) which uses the same Kjeldahl acid digest solution as that used for the nitrogen determination (see Discussion 7.7). The proper development of the yellow colour of the phosphovanadomolybdate complex is sensitive to the pH, which has an optimum value of 1.7. If spiky peaks shaped rather like an 'M', or an 'M' with an additional central peak, are obtained, it is probably caused by inaccurate adjustment of the sulphuric acid level to the 5-ml mark, or the diluted digest to the 10-ml mark in the digestion tube. The double peaks are caused by a difference in pH between the 50% sulphuric acid wash and the sample solutions.

Method 7.10a. Determination of total phosphorus in plant material by autoanalysis

Reagents.

- Acid-wash solution – wear PPE. Add 250 ml sulphuric acid (approximately 98% m/m H_2SO_4) to 250 ml water slowly with stirring. Allow to cool, then make up to 500 ml with water and carefully invert to mix.
- Ammonium hydroxide solution – using a fume cupboard, dilute 158 ml ammonia solution (approximately 35% m/m NH_3, 0.880 specific gravity) to 1 l.
- Ammonium molybdate solution – add 15 g of ammonium molybdate ((NH_4)$_6$$Mo_7O_{24}$.$4H_2O$) to about 900 ml warm water and stir to dissolve. Make up to 1 l in a volumetric flask and filter before use.
- Ammonium vanadate solution – *Caution*: wear PPE! Add 95.5 ml nitric acid (approximately 70% m/m HNO_3) to about 800 ml water in a 1-l volumetric flask. Add 0.5 g ammonium vanadate (NH_4VO_3), stopper and shake until dissolved, then make up to the mark and mix.
- Vanadate–molybdate reagent – mix the ammonium molybdate solution with the ammonium vanadate solution in the ratio 2:3 molybdate:vanadate. For optimum reproducibility, make up fresh each day just before use.
- Standard solutions – see Method 7.6a.

Procedure. The digestion procedure is given in Method 7.7a. The flow diagram is shown in Fig. 7.4.

To ensure adequate colour development, the flow passes through a double heating bath coil at 80°C. See also Note 2 in Method 7.7a. Procedure (autoanalysis). Switch on the autoanalyser modules and allow pumping of reagents for 30 min to flex the tubes. Carefully pour the sample digest solution from the digestion tube into the sample cup. Use 8.5-ml industrial polystyrene cups or similar size if more than one determination is required on the same sample solution. Load the first sampler tray in the sequence of low to high standards. It is recommended that the lowest standard is in duplicate, and the first peak rejected. After the highest standard, aspirate acid-wash solution for about 5 min to ensure the baseline is reached. Load further trays with about 32 sample cups followed by five standards. The remaining three spare cup positions can be used for higher standards or repeats. Again aspirate acid-wash solution to obtain a baseline between trays; this will enable a correction for any baseline drift to be made. If possible, keep samples of similar

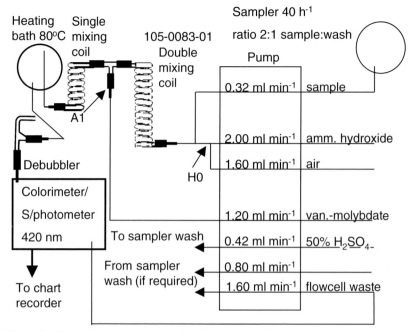

Fig. 7.4. Flow diagram for determination of phosphorus in plant digest solutions in 50% H_2SO_4.

analyte concentrations together to avoid interference between adjacent low and high peaks. The sampling rate is set at 40 h^{-1} with a sample:wash ratio of 2:1. When using a chart-recorder, record the sample number on every tenth peak and label the standard peaks; this makes reading the charts easier and enables the identification of problem peaks so that a repeat can be inserted.

Calculation. Draw a baseline under all the peaks by connecting the baselines obtained at the start, between the tray changes, and at the end. Draw a standard curve using a chart reader (see Chapter 1, 'Chart reader'). If exactly 0.1000 g sample was taken, then divide the concentration corresponding to the sample peak, in µg P ml^{-1}, by 100 to give the % total P in DM.

If y g sample was taken, then (µg P ml^{-1}) × 0.1/100y = % P in DM.

Discussion 7.11. Determination of total potassium in plant material and feeding stuffs

The official EC method for potassium in feeding stuffs may be downloaded from:

http://europa.eu.int/eur-lex/en/lif/dat/1971/en_371L0250.html

This is document 371L0250 (which is presented in the Official Journal No. L155, 12.07.71, pp. 13–37), and is the First Commission Directive 71/250/EEC of 15 June 1971. The determination of total potassium appears as Point 10 of the Annex. The sample ash is dissolved in HCl and the potassium content determined by flame photometry in the presence of caesium chloride and aluminium nitrate, which are included to largely eliminate interferences. The AOAC flame photometric method for plants is described by Isaac (1990, pp. 47–48). A flame photometric determination of potassium in plant solutions obtained by either dry or wet ashing is described in MAFF/ADAS (1986, pp. 191–192). This method says that there is no significant interference by other elements. We will describe (with Crown Copyright permission) the preparation of a sample solution of plant material by dry combustion as given in MAFF/ADAS (1986, pp. 8–9). The diluted Kjeldahl acid digest solution and standards from Method 7.7a may also be used for the flame photometric determination of potassium, but there are a couple of drawbacks to this convenience. First, because the sample solution is in 50% H_2SO_4, a total-consumption burner (where the sample capillary jet is situated in the burner nozzle) is unsuitable. This is because the atomized droplets of spray are incompletely vaporized by the flame, which causes a highly corrosive fall-out over the instrument and surrounding area (Faithfull, 1974). A premix burner with atomizer and spray chamber is satisfactory, however, the atomizer should be made of materials resistant to 50% sulphuric acid if nebulized directly from the sample cup.

Method 7.11a. Preparation of plant sample solution by dry combustion

Reagents.

- Hydrochloric acid, approximately 36% m/m HCl.
- Hydrochloric acid, approximately 6 M – mix equal volumes of hydrochloric acid, approximately 36% m/m HCl and water.

Procedure. Transfer 2 g dried sample, ground to 1 mm, into a silica basin, place in a cool muffle furnace, and increase the temperature to 500°C, which is maintained overnight. *Note*: if the sample is not oven-dried, take a separate sample for moisture determination. The ash should be whitish-grey. (If some carbon remains, cool the basin, moisten with water, dry at 102°C, then reheat at 500°C.) Cover with a watch glass and slightly displace to allow the addition of 10 ml of approximately 6 M HCl, avoiding loss of solution by effervescence. Remove and rinse the watch glass into the basin. Place the basin on a water-bath and evaporate the solution to dryness. When dry, continue heating for 1 h either on the water bath or in an oven at 102°C. Moisten the residue with 2 ml of hydrochloric acid, approximately 36% m/m HCl, cover the basin with a watch glass and gently boil for 2 min. Add approximately 10 ml of water and again boil. Remove the watch glass and rinse into the basin. Transfer the contents of the basin quantitatively into a 50-ml volumetric flask and dilute to the mark with water and mix. Filter through a 90-mm Whatman No. 541 filter paper, rejecting the first few millilitres, and retain the remainder for analysis in a polythene sample tube. Perform a blank extraction. Suitable determinations include Ca, K, Mg, Mn, P and Na.

Method 7.11b. Determination of potassium in plant material by flame photometry (dry ashing extract)

Standards

- Potassium stock standard, 1000 µg K ml^{-1} – dissolve 0.953 g KCl (dried for 1 h at 102°C) in water, add 1 ml hydrochloric acid, approximately 36% m/m HCl, dilute to 500 ml and add 1 drop of toluene.
- Potassium intermediate standard, 100 µg K ml^{-1} – pipette 25 ml of the potassium stock standard into a 250-ml volumetric flask, make up to the mark with water and mix.
- Potassium working standards, 0–50 µg K ml^{-1} – pipette 0, 10, 20, 30, 40 and 50 ml of the potassium intermediate standard into a series of 100-ml volumetric flasks, make up to the mark and mix.

Procedure. Switch on and set up the flame photometer according to the manufacturer's instructions. After a sufficient warm-up time, aspirate the 0 and 50 µg K ml^{-1} standards and adjust the zero and bring the maximum reading on-scale. Measure the other standards and construct a graph relating emission reading to concentration; check it is a straight line or smooth slight curve. Pipette 2.5 ml sample solution into a 100-ml volumetric flask, dilute to the mark with water and mix. Analyse the blank extract and the sample solutions in batches, and repeat the analysis of standards approximately every 10 min. If the output is on a chart-recorder, use a chart reader to assist the reading of the peaks (see Chapter 1, 'Chart reader'). Re-draw the standard curve if it changes during the analysis.

Calculation. Subtract the blank reading from the sample reading in µg K ml^{-1} read from the graph. This is equal to the g kg^{-1} of potassium in the sample. Divide the µg K ml^{-1} by 10 to obtain the % K in the sample. If the sample was not oven-dried, use the separate moisture determination to correct the result to K in DM.

Explanation: 2 g sample was dissolved in 50 ml solution (× 25 v/m dilution), which was further diluted by a factor of 40 (2.5 ml to 100 ml). Thus the 2-g sample would have been diluted to 50 ml × 40 = 2 l, which is 1 g sample l^{-1}. Therefore, 1 kg sample would be present in 10^3 l sample solution. If the measured concentration is y µg K ml^{-1}, this is equivalent to 10^3 y µg K l^{-1}, or 10^6 y µg K in 10^3 l, which is y g K in 10^3 l sample solution, which we have shown to contain 1 kg sample.

Method 7.11c. Determination of potassium in plant material by flame photometry (Kjeldahl acid digest)

Procedure. Switch on and set up the flame photometer according to the manufacturer's instructions. The standards (including a 50% sulphuric acid blank) and sample solutions in 50% sulphuric acid obtained from the Kjeldahl digest procedure 7.7a. may be aspirated directly into the flame photometer, but note the provisos in Discussion 7.10.

After a sufficient warm-up time, aspirate the 0 and 400 µg K ml^{-1} standards, and adjust the zero and bring the maximum reading on-scale. If the sensitivity of the instrument is too great, then dilute the standards and samples appropriately. Measure the other standards and construct a graph relating emission reading to concentration; check that it is a straight line or smooth slight curve. Analyse the sample solutions in batches, and repeat the analysis of standards approximately every 10 min. If the output is on a chart-recorder, use a chart reader to assist the reading of the peaks (see Chapter 1, 'Chart reader'). Re-draw the standard curve if it changes during the analysis.

Calculation. Divide the µg K ml^{-1} by 100 to obtain the % K in the sample, or by 10 to obtain the g K kg^{-1} in the sample. If the sample was not oven dried, use the separate moisture determination to correct the result to K in DM.

Discussion 7.12. Determination of starch by acid hydrolysis

The official EC method for starch in feeding stuffs is document 374L0203 (which is presented in the Official Journal No. L108, 22.04.74, pp. 7–24), and is the Fifth Commission Directive 74/203/EEC of 25 March 1974. The determination of starch appears as Annex I to the Directive. The AOAC gives an involved titrimetric method for starch in plants (Isaac, 1990, p. 60) and refers to older editions for starch in animal feed, including a direct hydrolysis method (Padmore, 1990, pp. 83–84). We will describe a method developed for starch in potatoes, but it may be used with other plant materials (Faithfull, 1990). See the discussion on the extraction procedure in Chapter 4, 'Fibre, lignin, cellulose, nitrogen-free extract and starch'. The starch is hydrolysed to glucose by heating with M HCl, which is then determined by autoanalysis using a colorimetric anthrone procedure. Correction factors are necessary to allow for the water content of the starch which remains after oven drying, for conversion of starch residue units of $C_6H_{10}O_5$ to the higher molecular mass units of glucose, $C_6H_{12}O_6$, and also for the conversion of any plant cell walls to chromogenic products.

Method 7.12a. Determination of starch in potatoes by hydrolysis and autoanalysis

Reagents (extraction).

- Benzoic acid, saturated solution
- Ethanol, 10% v/v
- Hydrochloric acid, 1 M
- Sodium hydroxide, 1 M

Procedure (extraction). Weigh 0.100 g of the dried (oven dried at 90°C, or freeze-dried) and milled (to 1 mm) sample into a 50-ml beaker; at the same time, weigh a sample for moisture determination at 110°C. A standard potato starch (Merck) sample should be included as a control. Add 10 ml of 10% ethanol and swirl to mix; allow to stand for 30 min to dissolve any dextrins, sugars and tannins, which could amount to about 12% by weight in DM. Centrifuge at 1500 g for 5 min, then wash the residue into a McCartney bottle, previously graduated at 15 ml, with 1 M HCl, and make up to the mark. Tightly screw on the cap, which has been lined with thick PTFE tape to prevent corrosion. Heat in an oven at 106°C for 40 min to hydrolyse the starch to glucose. Cool, and wash the contents with water into a 150-ml beaker containing 50 ml water. Adjust the pH to 3.0 ±0.2 dropwise with 1 M NaOH, transfer with beaker washings to a 100-ml volumetric flask, make up to the mark and mix. Allow any residue to settle, then pipette 10 ml into a 50-ml volumetric flask, and make up to the mark with saturated benzoic acid solution and mix. Analyse the sample by autoanalysis using the method given for water soluble carbohydrates (Method 7.14 below).

Calculation. The percentage starch in sample DM is given by the equation:

$$\% \text{ starch} = (F/200) \times 100/(100-W) \times 0.9 \times 100\%$$

where F = fructose concentration in sample solution (μg ml^{-1}); W = sample moisture content, and 0.9 is the correction factor to compensate for the conversion of $C_6H_{12}O_6$ back to the $C_6H_{10}O_5$ residue units in starch. A further factor of × 0.98 corrects for hydrolysis of potato cell walls. If required, the cell walls from the neutral detergent procedure may be used to find the correction factor for other plant species.

Discussion 7.13. Determination of trace elements in plants and feeds

The official EC method for Fe, Cu, Mn and Zn in feeding stuffs is document 378L0633 (which is presented in the Official Journal No. L206, 29.07.78, pp. 43–55), and is the Eighth Commission Directive 78/633/EEC of 15 June 1978. The determination of these trace elements appears as Point 3 of the Annex to the Directive. After ashing, the vegetable silica, which may retain some trace elements, is removed by evaporation with HF. This is too hazardous for educational purposes (untreated skin absorption could be fatal), and is not a feature of the AOAC methods. These methods are given for animal feed by Padmore (1990, pp. 84–85), and for plants by Isaac (1990, p. 42). Animal feeds (2–10 g) are ashed at 550°C for 4 h, heated with 3 M HCl, filtered and made up to 100 ml with 0.1–0.5 M HCl before analysis by AAS. Plant material (1 g) is ashed for 2 h at 500°C, 10 drops of water and 3–4 ml 50% v/v HNO$_3$ are added before evaporation at 100–120°C. The residue is again ashed for 1 h at 500°C and dissolved in 10 ml 50% v/v HCl, then transferred to a 50-ml volumetric flask. If Ca or Mg is to be determined, 10 ml of 5% La solution should be added to prevent interference by P in the air-acetylene flame. The

5% La solution is made up by dissolving 58.65 g La_2O_3 in 250 ml HCl, added slowly, and diluting to 1 l. Final determination is with AAS, and a modified method is also given for determination by ICP. MAFF/ADAS (1986, pp. 23–26) describe a wet-ashing technique using perchloric and nitric acids for digestion, and 6 M HCl for a subsequent boiling. See also the discussion in Chapter 4, 'Dry ashing', which suggests a possible loss of $FeCl_3$ at temperatures exceeding 450°C, which is the ashing temperature given in the method described below.

Method 7.13a. Determination of trace elements in plants and feeds

Reagents.
- Hydrochloric acid, approximately 36% m/m (for AAS)
- Hydrochloric acid, 0.1 M

Procedure. Weigh 2.0000 g of the dried and milled (to 1 mm) sample into a silica crucible and place in a cold muffle furnace with the chimney vent open, and allow to heat up to 450°C. Close the vent and maintain at this temperature overnight. Remove from the furnace and allow to cool, then add 15 drops HCl from a polythene Pasteur pipette, being careful to moisten all the sample. Using a fume cupboard, gently evaporate off all the HCl on a hotplate at moderate heat, then remove and cool. Dissolve the residue in 0.1 M HCl, and transfer quantitatively to a 10-ml volumetric flask. Make up trace element standards in 0.1 M HCl covering the expected ranges in the sample solutions and analyse by AAS (or ICP) according to the instrument manufacturer's instructions.

Calculation. The sample solution is of 2 g in 10 ml, therefore the concentrations in $\mu g \, ml^{-1}$ of the trace element should be multiplied by 5 to give the concentration in $\mu g \, g^{-1}$ of the trace element in the dried sample.

Method 7.14. Determination of water soluble carbohydrate by autoanalysis

See the discussion in Chapter 4 'Nitrate and water soluble carbohydrate'. *Note*: the extract obtained for determining nitrate in plants, which is also in saturated benzoic acid, may be used for the water soluble carbohydrate determination if diluted ×2.

Reagents.
- Anthrone reagent – dissolve 0.5 g anthrone in 500 ml sulphuric acid, 76% v/v. Prepare fresh daily and keep at 0–5°C.
- Benzoic acid, saturated solution.
- Sulphuric acid, 76% v/v – *Caution*: wear PPE. Carefully add 760 ml sulphuric acid (approximately 96% m/m H_2SO_4) to 330 ml water in a 2-l beaker standing in a sink containing cold water. Stir slowly with a glass

rod to mix and allow to cool. The solution will 'shrink' to approximately
1 l. Pour into a 1-l volumetric flask, make up to the mark, firmly stopper
and carefully invert to mix.

Standards.

- Stock fructose solution, 1000 µg ml⁻¹ of fructose – dissolve 1.0000 g fructose in saturated benzoic acid solution and make up to 1 l in a volumetric flask and mix.
- Working fructose standards, 10–300 µg ml⁻¹ of fructose – pipette 1, 5, 10,
  20, 25 and 30 ml of the stock fructose solution into a series of 100-ml volumetric flasks, make up to the mark and mix. This gives standards containing 10, 50, 100, 200, 250 and 300 µg ml⁻¹ of fructose.

Procedure. Weigh 0.1000 g freeze-dried plant material into a 250-ml wide-
mouth high-density polyethylene screw-cap bottle. At the same time, weigh
a sample for moisture determination. The square type bottles fit best the square
box of the reciprocating shaker. Add 100 ml saturated benzoic acid solution
and shake for 60 min. *Note*: For samples with < 25% WSC in DM, weigh
0.2 g sample. Filter through a Whatman No. 1, 15-cm filter paper, rejecting
the first few millilitres, and save in polythene capped sample tubes. The colour
is measured at 620 nm, and the flow diagram is given in Fig. 7.5.

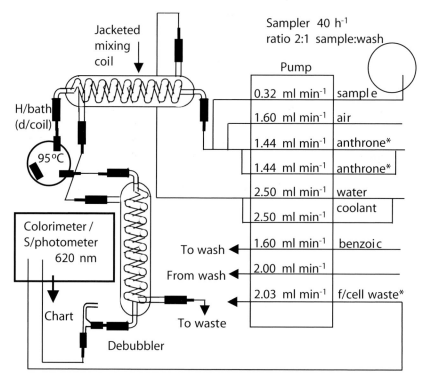

Fig. 7.5. Flow diagram for determination of water soluble carbohydrate.
        * Black acid-resistant tubing.

Note: the jacketed mixing coil has 14 turns (Bran+Luebbe Part No. 114-0222-01), and all the PVC sleeving tubing collars (ID 3.2 mm, ⅛ in) which connect the fittings under pressure should be wired on (a couple of turns of wire with the ends twisted together with pliers). This is because the viscous anthrone reagent causes considerable back-pressure, and if a fitting becomes disconnected, all the bubbles expand and force out a hot spray of 76% sulphuric acid. It may also be wise to tape a polythene sheet over the manifold in order to prevent serious injury from hot acid. Switch on the heating bath approximately 90 min beforehand to allow it to reach 95°C, then the rest of the modules for the autoanalysis. Pour the anthrone reagent into a conical flask, and then immerse in a beaker of crushed ice. Water coolant is a 2-l bottle of tap water, which has previously been refrigerated, and the wash solution is saturated benzoic acid. Commence pumping reagents and load the standards into a sample tray. After 20–30 min pumping, analyse at a rate of 40 samples per hour, adjusting the sensitivity of the detection-readout to bring the baseline and top standards on scale. If the samples are all low, the sensitivity should be increased and appropriate lower standards used. After a baseline, analyse a tray of samples plus standards, and follow with an adequate baseline before analysing the second tray of samples and standards. Conclude by aspirating the wash solution to obtain a final baseline. The sugar content of plants can vary greatly, so high peaks often obscure low ones, and repeats are often necessary. To reduce this inconvenience, try to group low and high sugar samples together. Check that the acid resistant pump tubes have not started to 'snake' due to stretching after an hour or so, as they are more prone to this than the ordinary PVC tubes; increase the tension to compensate for this, possibly at the start.

Calculation. Draw a baseline on the chart under all the sample peaks by connecting the baseline from aspirating wash at the start, between trays and at the end. Read the concentration of the sample solutions by comparing the peak heights of the samples with the standards using a chart reader (see Chapter 1, 'Chart reader'). Divide the concentration in $\mu g\ ml^{-1}$ of soluble carbohydrate in the sample solution by 10 to get the % water soluble carbohydrate in the freeze-dried sample. Multiply by 100/(100 – % moisture) to give the percentage water soluble carbohydrate in the sample DM.

8 The Analysis of Silage

Method 8.1. Determination of ammonium-N in silage

A distillation method is given in MAFF/ADAS (1986, pp. 168–169), but a selective ion electrode method will be described below.

Reagents.

- Alkali reagent (disodium EDTA, 0.1 M + sodium hydroxide, 2 M) – weigh 80 g NaOH and 37.2 g disodium EDTA into a beaker and dissolve in ammonia-free water. Transfer to a 1-l volumetric flask, and when cool make up to the mark and mix.
- Ammonia-free water – add about 6 ml sulphuric acid (approx. 98% m/m H_2SO_4) to 2 l water and distil off sufficient ammonia-free water, topping up the boiling container with more water as necessary.
- Stock standard solution, 1000 µg ml^{-1} of ammonium-N – dissolve 0.955 g NH_4Cl (previously dried at 102°C for 1 h) in ammonia-free water and make up to 250 ml and mix.
- Working standard solutions, 0–200 µg ml^{-1} of ammonium-N – pipette 0, 1, 2.5, 5, 10, 15 and 20 ml of the stock standard into a series of 100-ml volumetric flasks to obtain standards of 0, 10, 25, 50, 100, 150 and 200 µg ml^{-1} of ammonium-N. *Note*: solutions of standards and samples should be equilibrated to the same (room) temperature before measurement.

Procedure. Weigh approx. 20 g fresh silage into a 250-ml wide-mouth high-density polyethylene screw-cap bottle. Add 100 ml ammonia-free water and shake for 1 h. Filter through a 150-mm Whatman No. 1 filter paper. Pipette 20 ml of the extract into a 50-ml beaker containing a magnetic stirring bar and add 2 ml of the alkali reagent. Immediately stir for exactly 1 min and insert the ammonia electrode. (*Note*: if the electrode has been stored in 0.1 M NH_4Cl solution, it should be rinsed thoroughly before use.) Very low readings may take several minutes to stabilize. Repeat the determination using standards made up using ammonium chloride. Compare sample readings with the standard curve to determine the concentration of ammonium-N in the extract solutions. *Note*: the standard curve is prepared by plotting on semi-log graph paper (e.g. Chartwell Graph Data Ref. 5231), which is Log 3 Cycles for the shorter x-axis (concentration) and mm, ½ and 1 cm for the y-axis (mV). The greater the concentration, the greater the negativity in mV; it may therefore be better for the y-axis to go from zero at the bottom to −200 mV at the top of the axis. The sensitivity is quoted as about 56 ±3 mV per decade (= one log cycle, e.g. from 10 to 100 µg ml^{-1} ammonium-N) at 25°C. The graph is nearly a straight line, or slightly concave. A plot on ordinary graph paper is very convex and difficult to read accurately.

Calculation. Let the µg ml^{-1} ammonium-N be y, then this is 100y µg 100 ml^{-1}, or y/10 mg 100 ml^{-1} of extract solution per 20 g fresh silage. Therefore the weight of ammonium-N per 100 g fresh silage is: (100/20) × y/10 mg, or 0.5y/1000 % ammonium-N, which is y/2000 % ammonium-N.

This must next be expressed in terms of DM. If the moisture content is m, then this becomes:

y/2000 × (100/100 − m) % ammonium-N in DM.

Finally, this is conventionally expressed as a percentage of the total N % in DM. The final expression for ammonium-N as a percentage of total-N in DM is therefore:

y/2000 × (100/100 − m) × (100/total-N), which reduces to
5y/(100 − m) (total-N%).

A typical sample gave a reading of −148 mV, equivalent to 45.5 µg ml^{-1} ammonium-N. The moisture content was 66.46%, and the total-N was 1.83% in DM. Substitution in the above equation gives: 45.5 × 5/(100 − 66.46)(1.83) = 227.5/61.378 = 3.71% ammonium-N as a percentage of total-N in DM.

Interpretation. The ammonium-N expressed as a percentage of total-N should not exceed about 11% for a good fermentation. Levels over 15% reduce palatability and can reduce voluntary intake.

Method 8.2. Determination of moisture in silage

See Chapter 4, 'Water content in silage'. The subjection of silage to oven drying causes loss of volatile components as well as water, resulting in overes-
timated moisture content. One way of minimizing this effect is to distil the
fresh silage in the presence of toluene. Titration of the acidic distillate with
0.1 M NaOH enables a correction to be made for the volume of the volatile
acids. A suitable method is given in MAFF/ADAS (1986, pp. 85–87) where
about 70 g silage plus 400 ml toluene is heated in a 1-l round-bottomed flask,
the distillate collecting in a specially made Dean and Stark receiver. The
modified procedure is described below and in Faithfull (1998); it uses only
10 g samples, which are distilled with 100 ml toluene from a 250-ml flask.
The aqueous distillate is collected in a standard Quickfit® Dean and Stark
receiver fitted with a Rotaflo® stopcock permitting easy release of the water
before titration. The apparatus is shown in Fig. 8.1, and captions refer to
Quickfit® Part Nos.

Reagents.

- Ethanol
- Phenolphthalein indicator, 0.1% (m/v) – dissolve 0.1 g phenolphthalein in
  100 ml 95% ethanol.

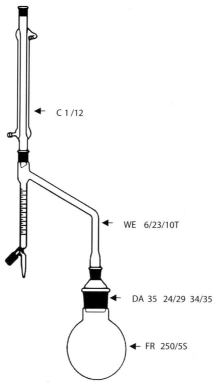

C 1/12

WE 6/23/10T

DA 35 24/29 34/35

FR 250/5S

Fig. 8.1. Modified Dean and Stark apparatus for small silage samples.

- Sodium hydroxide, 0.05 M
- Toluene

Procedure.

DISTILLATION. Weigh 10.0 g fresh or frozen silage into a beaker. Immediately replace the rest of the sample into the deep-freeze to prevent loss of volatiles. Transfer to a 250-ml round-bottomed flask, supported on a cork or rubber ring. Add 100 ml toluene, place in a heating mantle of the correct size, and connect to the 10-ml Dean and Stark pattern receiver/condenser assembly. Turn the energy controller on full to bring to boiling, then adjust to give a steady boil. After 20 min, and then after 5-min intervals, record the volume of aqueous phase in the receiver until two identical values are obtained. Dislodge any droplets adhering to walls with a glass rod or gentle rocking. Switch off the heater and allow the receiver to cool before recording the final volume. Put a black card behind the receiver to show the meniscus clearly. Run off most of the aqueous phase into a 50-ml beaker. Pipette 5 ml of the aqueous phase into a 50-ml conical flask for titration to determine the acid content.

TITRATION. Add 20 ml ethanol to the 5 ml aqueous phase and 5 drops phenolphthalein indicator. Pipette into the flask 10 ml 0.05 M sodium hydroxide and titrate with the same solution from a 10-ml burette after noting the initial volume. *Note*: Read the burette to 0.02 ml accuracy. Titrate to the first permanent pink colour; this fades due to CO_2 from the air, so don't delay. Add the 10 ml initially pipetted into the flask to the volume delivered from the burette to obtain the total titre (probably in the range 13–18 ml).

Calculation. The dry matter (g kg^{-1}) is calculated according to the formula:

$$DM = 1000 - \frac{998 \times V}{m}\left(1 - \frac{ft}{10}\right)$$

DM = dry matter g kg^{-1} m = mass of silage sample (g)
V = volume of aqueous distillate f = factor (0.00555)
t = titre   998 = acid:water density correction

Microsoft® Excel Program. It is usual to use PC software such as Microsoft Excel to perform repetitive calculations such as the above. This enables a tabular printout and the generation of charts. The layout of the data sheet is shown in Fig. 8.2. The first four columns of data are entered directly. The fifth column is a correction of the titration using the actual strength of the sodium hydroxide used compared to what the result would have been if the concentration had been exactly 0.05 M. Thus, the equation to be entered in the cell to convert the previous column's value is in our case:

=G2*(0.048/0.05)

where G2 is the previous column's cell, 0.048 is the molarity of the acid used, and 0.05 is the molarity to which the titre is being corrected.

Sample No. Pit 1/	Mass of Silage(M)	Volume of Water (V)	Vol. ~.05M NaOH	Vol. .05M NaOH (T)	DM of Silage g/kg⁻¹	Moisture in Silage %
1	10.00	7.70	21.30	20.45	240.26	75.97
2	10.00	7.25	19.02	18.26	283.78	71.62
3	10.00	7.30	19.48	18.70	279.02	72.10
4	10.00	7.25	19.76	18.97	284.07	71.59
5	10.00	7.25	22.50	21.60	285.12	71.49
6	10.00	7.40	17.20	16.51	268.25	73.18
7	10.00	7.70	18.70	17.95	239.20	76.08
8	10.00	7.50	14.26	13.69	257.19	74.28
9	10.00	7.60	14.70	14.11	247.46	75.25
10	10.00	7.35	16.64	15.97	272.97	72.70
11	10.00	7.20	15.68	15.05	287.44	71.26
12	10.00	7.60	18.68	17.93	249.07	75.09
13	10.00	7.90	17.66	16.95	219.00	78.10
14	10.00	7.70	15.66	15.03	237.95	76.20
15	10.00	7.50	16.92	16.24	258.25	74.18
16	10.00	7.15	17.94	17.22	293.25	70.67
17	10.00	7.50	15.16	14.55	257.55	74.25
18	10.00	7.45	14.64	14.05	262.29	73.77
19	10.00	7.65	14.72	14.13	242.52	75.75
20	10.00	7.80	16.38	15.72	228.35	77.16
21	10.00	7.45	15.20	14.59	262.51	73.75
22	10.00	7.75	16.08	15.44	233.18	76.68
23	10.00	7.60	13.54	13.00	246.99	75.30
24	10.00	7.30	15.00	14.40	277.28	72.27

PenglaisFirst Cut, First Sampling

DM (dry matter g/kg)

		38 cm	76 cm	114 cm	152 cm	190 cm	228 cm
Sampling	1	240.26	283.78	279.02	284.07	285.12	268.25
Position	2	239.20	257.19	247.46	272.97	287.44	249.07
	3	219.00	237.95	258.25	293.25	257.55	262.79
	4	242.52	228.35	262.51	233.18	246.99	277.28

Moisture%

		38 cm	76 cm	114 cm	152 cm	190 cm	228 cm
Sampling	1	75.97	71.62	72.10	71.59	71.49	73.18
Position	2	76.08	74.28	75.25	72.70	71.26	75.09
	3	78.10	76.20	74.18	70.67	74.25	73.77
	4	75.75	77.16	73.75	76.68	75.30	72.27
		305.90	299.26	295.28	291.64	292.30	294.31

		38 cm	76 cm	114 cm	152 cm	190 cm	228 cm
Av. Moisture%		76.48	74.82	73.82	72.91	73.08	73.58

Fig. 8.2. Microsoft® Excel data sheet for moisture in silage from the Penglais third-cut, first sampling.

The next column converts the various data to dry matter of silage g kg⁻¹. The formula to be entered in the first cell of this column is:

=1000–((998*E2)/10)*(1–((0.00555*I2)/10))

where E2 is the cell referring to the volume of aqueous distillate and I2 is the corrected volume of 0.05 M sodium hydroxide.

There are also other tables on the data sheet that allow the easier production of charts. They refer to DM and moisture at each depth and sampling position, also the average moisture for each depth.

Interpretation. The interpretation of the effects of moisture in silage is notoriously difficult. Moisture is essential for the proliferation of desirable microorganisms, but an excess will encourage the growth of undesirable types (Woolford, 1984). The DM range most favourable to the silage fermentation has been suggested as 200–250 g kg⁻¹ and the optimum about 240 g kg⁻¹.

Method 8.3. Determination of pH in silage

This is a modified version of the method of MAFF/ADAS (1986, pp. 102–103).

Procedure. Weigh approx. 50 g fresh silage into a 500-ml beaker and add 125 ml water from a measuring cylinder. Using the base of a suitably sized measuring cylinder as a piston, pump the silage up and down to ensure thorough mixing with the water. Allow to stand for exactly 1 h with occasional mixing as described. While standing, adjust the pH meter using pH 7.0 and pH 4.0 buffer solutions. Decant into a smaller beaker and measure the pH of the silage extract.

Interpretation. The optimum pH is 3.8–4.2. Values of pH < 7.0 indicate acidity, in this case mainly from the lactic acid produced by the fermentation of 'well preserved silage'.

Discussion 8.4. Determination of volatile fatty acids (VFAs) in silage

We still await a routine method for the rapid analysis of large batches of silage samples for VFAs. Although initially expensive, the use of NIRS on fresh silages would be ideal, but sufficiently robust equations for components other than lactic, acetic, propionic and butyric acids are not yet widely available, although it is reported that commercial services are offering VFA analysis by NIR. The abstract of the poster by Deaville and Givens (1996) is apparently the only published data in this area (D.I. Givens, ADAS, Stratford-on Avon, 2001, personal communication), which suggests further work is needed to improve the accuracy of prediction, especially for acetic and propionic acids. That leaves gas chromatography (GC) and high performance liquid chromatography as the two methods most commonly employed, although automatic titration methods also exist. The chromatographic methods have traditionally required that the acids undergo a time-consuming derivatization step (Jones and Kay, 1976; MAFF/ADAS 1986, pp. 235–239) converting them to esters, which are more easily separated and detected. For about 20 years, procedures have been developed allowing the direct injection of silage juice, or aqueous extracts. The analysis of synthetic mixtures of the VFAs usually give clearly separated peaks, but when the actual silage extract is injected, there are several problems. For example, in GC, lactic acid tends to give a large broad peak that obscures the following small peaks of any *n*-valeric or iso- and *n*-caproic acids. In HPLC, there are so many different organic compounds in the silage extract with each producing a peak on the chromatogram, that it is difficult to identify a peak with any degree of certainty. Even knowing the exact elution time of the pure acid is of no real benefit when analysing the silage extract, because an increase in the column temperature or in the acidity of the eluant has been found to cause the peak previously attributed to just one VFA to be resolved into two. There can thus be two or more peaks eluting simultaneously, which leads to an erroneously high estimate of concentration for that particular VFA. Succinic acid, for example, elutes at only about 18 s before lactic acid in HPLC analysis using a Bio-Rad Aminex HPX 87 H ion-moderated partition column, and may well overlap to varying degrees. It may be advantageous to analyse some components, such as ethanol, by GC, and others, such as lactic acid, by HPLC. There are numerous columns available for both GC (packed and capillary) and HPLC, each with special instructions from the manufacturer.

In this case, therefore, an actual analytical procedure will not be recommended, but references to some published methods and some guidelines will be given.

Gas chromatography of VFAs

A GC method using direct injection of silage juice extract has been proposed by Fussell and McCalley (1987) using a Carbopack B-DA/4% Carbowax 20M (80–120 mesh) column. This improved on the tailing lactic acid peaks of Playne (1985) who used a Chromosorb 101 (80–100 mesh) column. A column capable of derivatizing *in situ* was suggested by Suzuki and Lund (1980). It consisted of poly(ethylene glycol phthalic acid ester) coated on a solid terephthalic acid support, and sharpened up the lactic acid peak. Galletti and Piccaglia (1983) used a Porapak QS (80–100 mesh) column to separate lactic acid and C_2–C_6 VFAs in 10 min using acidified portions of aqueous maize silage extract.

We obtained the best results with the Carbopack B-DA/4% Carbowax 20M, 80–120 mesh Supelco column, 2000 × 2 mm. It was initially conditioned for 21 h at 245°C, but the normal running temperature is 175°C. The injector/detector temperature is 200°C and a flame ionization detector is used. A glass sleeve is fitted to the injector and the glass wool plug removed from the column inlet. The carrier gas is nitrogen with a flowrate of 40 ml min^{-1} at 310 kN m^{-2}. The sample solution (9 ml) is mixed with 1 ml of pivalic acid solution (1.6% m/v) as internal standard. Then 1 ml of this solution is mixed with 1 ml 0.3 M oxalic acid solution and 3 ml deionized water before injecting 1 μl into the septum.

A GC chromatogram of a mixture of known VFAs, lactic acid and pivalic acid (internal standard) is shown in Fig. 8.3, and a chromatogram of a silage juice extract with pivalic acid internal standard is shown in Fig. 8.4.

HPLC of silage VFAs

Careful work at the Scottish Agricultural College (www.sac.ac.uk) by Rooke *et al.* (1990) and Salawu *et al.* (1997) has enabled the successful analysis of silage VFAs using the Bio-Rad Aminex HPX-87H ion-moderated partition column and a refractive index detector that enables ethanol in addition to the VFAs to be detected. They also successfully used an equivalent Supelco Supelcogel C-610H 300 × 7.7 mm ID column. In either case, a guard column is used (Bio-Rad Cation H$^+$ Cat. No. 125-0129). Other workers (Kubadinow, 1982; Canale *et al.*, 1984; Siegfried *et al.*, 1984) have also used the Aminex HPX-87H 300 × 7.8 mm column with an Aminex HPX-85H 40 × 4.6 mm guard column and UV detection at 210 nm.

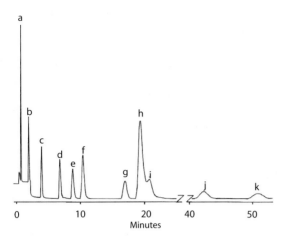

Fig. 8.3. GC chromatogram of mixed silage juice standards using a Carbopack B-DA column. Identity and concentrations (before diluting 4:1 standard:0.3 M oxalic acid): a, ethanol, 1 mg ml^{-1}; b, acetic acid, 1.25 mg ml^{-1}; c, propionic acid, 0.25 mg ml^{-1}; d, isobutyric acid, 0.25 mg ml^{-1}; e, *n*-butyric acid, 0.25 mg ml^{-1}; f, pivalic acid (internal standard), 0.4 mg ml^{-1}; g, isovaleric acid, 0.25 mg ml^{-1}; h, lactic acid, 10 mg ml^{-1}; i, *n*-valeric acid, 0.25 mg ml^{-1}; j, isocaproic acid, 0.25 mg ml^{-1}; k, *n*-caproic acid, 0.25 mg ml^{-1}.

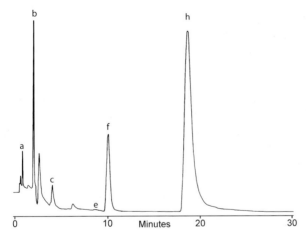

Fig. 8.4. GC chromatogram of a third-cut silage juice using a Carbopack B-DA column. Identity: a, ethanol; b, acetic acid; c, propionic acid ; e, *n*-butyric acid; f, pivalic acid (internal standard), 0.32 mg ml^{-1} in injected solution; h, lactic acid.

We experimented firstly with a Phenomenex Rezex ROA-Organic Acid 300 × 7.8 mm column with a Rezex Organic Acid 50 × 7.8 mm guard column; the mobile phase was 0.013 N (0.0065 M) H$_2$SO$_4$. UV detection was at 215 nm, the flowrate 0.6 ml min^{-1}, column temperature 35°C and injection volume 20 µl. The internal standard was 2-ethylbutyric acid. The lactic acid peak was preceded by a partially overlapping, probably succinic acid peak; the

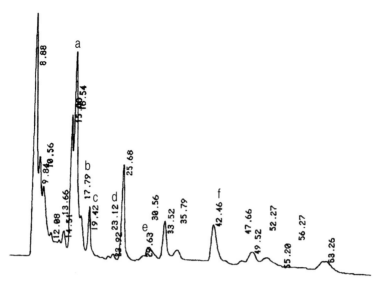

Fig. 8.5. HPLC chromatogram of a first-cut silage juice using a Rezex ROA-organic acid column. Identity: a, lactic acid; b, formic acid; c, acetic acid; d, propionic acid; e, *n*-butyric acid; f, 2-ethylbutyric acid (internal standard).

two peaks coincided at 45°C. A typical chromatogram is shown in Fig. 8.5, with elution times in minutes added by the integrator-printer and positively identified acids (acetic, formic, propionic, *n*-butyric and lactic), indicated by letters over the elution times.

Second, we experimented with a Spherisorb C_8 5µ column using 0.2 M H_3PO_4 as solvent at a flowrate of 0.6 ml min^{-1}. A typical chromatogram is shown in Fig. 8.6.

The peak at 17.34 min is 50 µg ml^{-1} mesaconic acid which was added as a possible internal standard, but it proved unsuitable because of an unknown peak eluting at 16.78 minutes seen when analysing silage juice without added internal standard; oxalic acid suffered from the same problem. We still need a suitable internal standard for this column, thus further use was suspended. Other failed compounds included: adipic acid, fumaric acid, D-glucuronic acid, glutaric acid, glycolic acid, 3-hydroxybutanone, itaconic acid, malic acid, maleic acid, malonic acid, pimelic acid and succinic acid.

Extraction procedure

We have found that whatever separation/detection method is used, the extraction procedure used for obtaining the silage juice should be standardized. This is because slightly different amounts of the various components will be extracted from the fibrous silage material depending on whether it is homogenized, shaken or compressed. A compression method is given in 8.4. below. Even the strength of torque applied in the compression method can affect the recovery of VFAs. Thus the recommended torque of 8 Nm recovers on

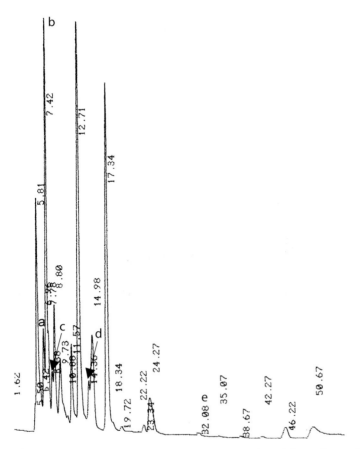

Fig. 8.6. HPLC chromatogram of a first-cut silage juice using a Spherisorb C_8 5µ column. Identity: a, formic acid; b, lactic acid; c, acetic acid; d, propionic acid; e, *n*-butyric acid. The peak at 17.34 min is 50 µg ml⁻¹ mesaconic acid added as a possible internal standard, which proved unsuitable because of an unknown peak eluting at 16.78 min obtained using silage juice without added internal standard.

average 12% more of the various VFAs and lactic acid than a torque of 5.4 Nm, which is easier to apply. This is to be expected if, as is likely, the concentration of VFAs is not homogeneous throughout the sample, and thus any juice held in the deeper interstices will be expressed by the greater torque.

Internal standard

It is good practice to include an internal standard, which should not be degraded in any way, such as by being heated in a gas chromatograph. It should also give a peak that does not overlap any of the peaks of the sample being determined. Acids and ketones should be suitable, but aldehydes are not recommended. Pivalic acid is often used for this purpose in the GC analysis of VFAs. Although variations in the sensitivity of the determination can be corrected by measuring the peak area of a constant amount of pivalic acid included with the sample, it has its limitations. Repetitive injections of the same sample of silage juice may show that any variation in sensitivity between injections is not always equal for all the peaks. However, the use of an internal standard reduces the error and improves the precision. With HPLC of silage extracts, a widely used internal standard is 2-ethylbutyric acid.

Method 8.4. Extraction method for obtaining silage juice for analysis for VFAs

Equipment.

- Extractor – the extractor is shown in Fig. 8.7, and the means of attaching the small torque wrench in Fig. 8.8. Ideally it should be constructed from stainless steel, but we only used a stainless steel threaded rod or studding (10 × ⅝ in, 11 t.p.i., or 254 × 15.9 mm), and the rest was machined from brass. The silage juice should not remain in contact with the brass for long, however, because the brass will be attacked by the acids. The dimensions

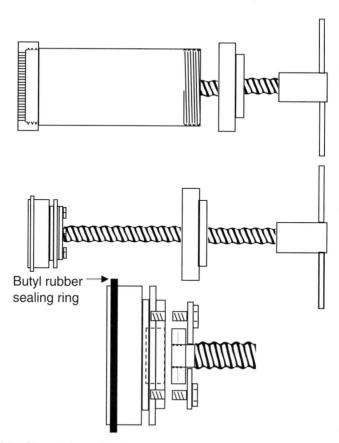

Butyl rubber
sealing ring

Fig. 8.7. Silage juice extractor.

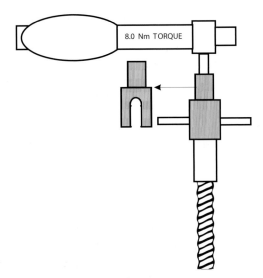

Fig. 8.8. Attachment of torque wrench to silage juice extractor by means of a socket.

of the cylindrical body are 2 in ID × 2½ in O.D. × 9 in long (approx. 50 mm ID × 63.5 OD × 230 mm long). The piston or plunger contains a groove, which is fitted with a butyl rubber (oil-seal) ring. The top and bottom of the cylinder are threaded for easy removal and cleaning. The bottom measures 2¾ in (70 mm) diameter and is 1 in (25 mm) in depth. It is perforated with 75 holes of 3 mm diameter, and 20 holes of 2.5 mm diameter. The top is the same diameter as the bottom, but is 1¼ in (32 mm) in depth to allow sufficient thickness for the portion threaded for the studding.

* Torque wrench and socket – a small size torque wrench capable of being set at 8.0 Nm, and a socket having two slots milled opposite to each other to fit over the ⅜ in (9.5 mm) diameter T-bar at the end of the threaded rod.

Procedure. Load about 40 ml of fresh (or thawed) silage into the extractor and screw on the top, with the plunger piston located at the top end. Attach the torque wrench, fitted with the socket, over the T-bar. Slowly increase the torque to about 5.4 Nm over 0.5–1 min and collect the juice in a beaker. About 70–80% of the silage moisture content will be obtained. Then increase the torque to 8.0 Nm, allow the few extra millilitres of juice to drain into the beaker, then re-apply the torque to obtain the remaining drops. Next centrifuge the juice in polypropylene tubes at 20,000 *g* for 10 min at a temperature of approx 4°C to prevent loss of volatiles. Analyse immediately, but if this is not possible, store the extracts in capped polythene sample tubes in a deep freeze. On thawing, the solutions may appear cloudy, and a fine brown residue will inevitably be deposited. Therefore, filter a sufficient quantity of the solutions using a 25 mm × 0.45 μm polyether sulphone (PES) membrane syringe filter (e.g. Syrtec 0.45 µm PES sterile, Cat. No. 8670180142, Techmate: www.techmate.co.uk) for the analysis, which should be carried out immediately. *Note*: PES is not the same as polysulphone (PSul). A porosity as fine as 0.45 µm is essential to remove particles that could seriously reduce the life of an HPLC column or guard cartridge.

9 Near Infrared Spectroscopy

Practical instructions regarding this method of analysis are beyond the scope of this book, however, some details will be given to provide a background for those wishing to make use of commercial services. Some material was originally published in Faithfull (1996).

What exactly is NIRS? The near infrared region of the electromagnetic spectrum is a region having a range of wavelengths slightly longer than visible light but not as long as microwaves or the longer radiowaves. A beam of light from a quartz-iodine lamp is shone on to the sample and the spectrum of the reflected light is analysed by a spectrometer. Some wavelengths of the light beam will have been reduced in intensity because of absorption by certain vibrating molecular bonds (in particular C–H, O–H and N–H). It is not the fundamental vibration that is involved in this region but overtones or combinations of the fundamentals. The spectrum must be sampled at about 700 data points so that the subsequent data processing can unscramble the interacting spectral peaks and relate them to concentration of a substance of interest in the original sample. This process is largely mathematical and statistical and far removed from wet chemistry where, for example, the actual fibre is extracted from the sample and weighed. A computer system is required in order to process the data produced by the spectrometer.

Prediction of Metabolizable Energy (ME)

Instead of predicting the ME value of silage from the MADF measurement using a regression equation (see Chapter 4, 'Fibre, lignin, cellulose, nitrogen-free extract and starch', ADAS is recommending the discontinuation of the above method in favour of one using NIRS methodology (Barber *et al.*, 1990). The procedure involves three stages:

1. Prediction of organic matter digestibility (OMD) from NIRS data;
2. Conversion of OMD to DOMD (digestibility of organic matter in oven dried matter) using ash values, and then correction to true dry matter basis (allowing for volatiles) to give DOMDc;
3. Use the equation ME = DOMDc × 0.16.

This procedure, however, has the built-in problems of the difficulties in standardizing the NIRS methodology and then using two subsequent regressions. The advantage of NIRS over MADF has been summarized by Offer (1993). Basically, different analytical methods and regression equations have produced various ME values from the same silage sample. ADAS and other advisors, including the animal feed trade (UKASTA – United Kingdom Agricultural Supply Trade Association), met and agreed to the newly proposed method in an attempt to remedy the situation. The reason for preferring the new method is that extensive *in vivo* feeding trials of 200 silages showed that MADF gave poor correlation with OMD whether carried out by wet chemistry or predicted by NIR. OMD was best predicted by NIR using an eight-term linear regression. This was, however, very sensitive to interference from moisture and particle size effects. Further mathematical treatment to remedy this defect led to the recommended 76-term partial mean square model regression equation. The resulting ME values, however, were found to be 0.4 MJ kg^{-1} (or 1.2 MJ kg^{-1} CDM – corrected dry matter) greater than previous predictions using DOMD$_o$ × 0.16. The reduction of concentrate allowances to dairy cows by 20% as suggested by these figures was judged unacceptable and the feed allowance should therefore be adjusted to allow for this underestimate. The NIR–OMD–ME route, however, remains the most precise of the available methods.

Applications of NIR

Hay has been analysed by NIR for crude protein, acid detergent fibre, dry matter, lignin and IVDMD, rapeseed for oil and water; and spring field beans for N to name but a few applications. Most macroinorganic constituents of peaty soil can be determined, and moulds have been measured in hay, tall fescue and barley (Malley and Nilsson, 1995). A short bibliography is given below.

Compost (C and N)	Suehara *et al.* (2001)
Food analysis	Osborne and Fearn (1986)
Forage analysis	Parnell and White (1983); Shenk *et al.* (1992)
Manures (ammonium-N, moisture, total C and N)	Reeves and Van Kessel (2000)
Ryegrass (ADF, ash, crude fibre, crude protein, IVOMD, lignin, NDF)	Berardo (1992)

Silage (DM)	Baker and Givens (1992)
Silage (DM, crude protein, NDF, Cellulase OMD)	de la Roza and Martinez (1992)
Silage (DOMD)	Hellämäki (1992)
Silage (VFAs)	Deaville and Givens (1996)
Soil (biological activity)	Reeves *et al.* (2000)
Soil (OM, humus)	Krischenko *et al.* (1992)
Soil	Reeves and McCarty (2001)
Spectra	Williams and Norris (1987); Hildrum *et al.* (1992)
Sulphur in plants	Batten and Blakeney (1992)
Theory	Stark and Luchter (1991)
Wheat	Barton *et al.* (2000)

The *Journal of Near Infrared Spectroscopy* has a website at:
  http://www.nirpublications.com/
and abstracts of journal articles are available at:
  http://www.nirpublications.com/jnirabst.html
One site describing NIR equipment may be viewed at:
  http://www.foss-nirsystems.com/

Interpreting NIR Spectra

Unlike the relatively straightforward infrared spectra, which arise mainly from fundamental molecular vibrations and rotations, and where molecular components can be readily ascribed to the 'fingerprint' of peaks, the visual interpretation of NIR spectra is virtually impossible and speculative at best.

  There are two approaches to interpreting spectra (Bonanno *et al.*, 1992, p. 19):

1. Try and relate observed bands and peaks with known absorbing functional groups or chemical compounds;
2. Take a chemometric approach: ignore the question as to what causes the peak, and select the absorbing wavelength on an empirical basis to give the best correlation with traditional chemical analyses.

The difficulty with the first approach can be seen in the case of chloroform ($CHCl_3$), which has only one absorbing CH group, yet possesses no less than 62 possible combination bands (Kaye, 1954). The presence of many C–H bonds in different molecular locations in organic macromolecules, such as the ligno-celluloses, will lead to vast numbers of absorption frequencies. Most of the absorbing frequencies derive from overtones and combinations of fundamental vibrations involving hydrogenic stretching modes (Osborne and Fearn, 1986, p. 29). There is also extensive overlapping and perturbation of the NIR absorption bands. It must be noted, however, that only combination bands arising from two different vibrational modes of the same functional group and having the same symmetry are allowable (Bonanno *et al.*, 1992,

p. 22). The CH stretching absorber is also present in proteins, oils and starch as well as cellulose, complicating the spectral interpretation. A similar interpretational problem exists with OH absorbers. These are present in simple sugars, but many of the same bands also appear in starch and cellulose (Murray and Williams, 1987). Williams (1991) has stated that 'At any wavelength area between 750 and 2500 nm, there is a multiplicity of absorbers, all of which may contribute to the spectrum of a commodity. For example, in the 2100 nm area, nearly 20 absorbers, including 2nd and 3rd overtones, can be identified, and the assignment of the wavelength to any particular absorber becomes rather specious.' The second, empirical approach, however, lacks a sound physico-chemical basis, but can be made to work under the right conditions. Irrespective of the compounds causing the overlapping spectral bands, it is the shape, that is the rate of change in slope with respect to wavelength, that conveys compositional information (Deaville and Baker, 1993). Different types of organic composite substances thus possess a 'fingerprint', and the relative position and magnitude of the peaks can be interpreted to yield information on the composition and relative amounts of substances present.

There is a warning from Shenk *et al.* that 'simply running data through the latest mathematical algorithm will result in nothing interpretable and is only pseudoscience.' Nevertheless, Givens (1993) claims that in most cases, these equations have been shown to provide a better prediction of forage digestibility, for example, than laboratory procedures.

Interferences

Interfering effects must be minimized. This is achieved by mathematical transformations, and two such accepted procedures are, first, the SNV (standard normal variate) transformation which standardizes the variance of the spectrum to unity with a mean of zero. This minimizes particle size effects and baseline drift. Second, the de-trending (D) transformation, which removes curvilinearity of the spectrum by use of a second-order polynomial correction (Barnes *et al.*, 1989). Another procedure uses a repeatability file to remove interference from moisture on spectra from similar samples analysed over a long period (Deaville and Baker, 1993). The main spectral interference in the NIR of agricultural materials arises from the presence of water, which possesses a strong absorption at 1450 and 1930 nm. It is present even in materials dried at 100°C. Water forms strong hydrogen bonds with cellulose and other materials containing OH or NH groups. This makes the water difficult to remove and affects the absorption, causing a shift of up to 50 nm to longer wavelengths and band broadening (Shenk *et al.*, 1992). The presence of moisture has two other quantitative effects: it will affect the height of the peaks and hence the estimated concentration of components; also, if the concentration is reported as percentage in DM, it will cause an underestimate of the true concentration.

To test whether it was practicable to relate any areas of the observed NIR spectrum to chemical components, various species of grass and weeds were fed to sheep, and the strained rumen liquor and its protozoal and bacterial fractions were freeze-dried and the NIR spectra obtained. A typical spectrum for the ryegrass diet is shown in Fig. 9.1, the data being subjected to a standard normal variate and detrend procedure (SNV-D) before plotting against wavelength.

It proved impossible to relate the peak shapes and heights to a particular component and its concentration. The variation in peak shapes and heights are also extremely small, and visually difficult to discern. There are parts of the spectrum that are claimed to be areas of special biological significance. For the spectra to be meaningful, however, they must have been derived from a statistically significant number of similar samples to permit the use of correlation transform techniques to identify the analytically useful wavelength regions. These are not always available in investigative work. The margin between a valid statistical interpretation and doubtful conjecture is therefore very small in the case of the average analytical laboratory with only a limited number of samples with which to set up the algorithms. Only with organizations operating on a large scale does NIR gain in credibility. The interpretation of large numbers of NIR spectral data obtained from ADAS in relation to predicted and actual feeding value was the subject of a PhD thesis by Field (1995).

The capabilities of NIR to provide rapid analytical results that are acceptably accurate and precise, providing samples are similar to those used in the calibration procedure, are well established. It is still common practice, however, to analyse one sample in every 20 by wet chemical methods just to be sure that the calibration remains valid.

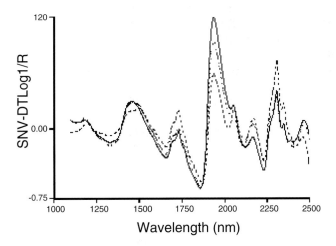

Fig. 9.1. The NIR spectra from strained and freeze-dried whole rumen liquor from sheep fed on ryegrass (solid line), and the bacterial (dotted line) and protozoal (dashed line) fractions. The standard normal variate and detrended data (SNV-DT) is plotted vs. wavelength. Log 1/R = − log (Reflectance) and is equivalent to absorbance.

10 Methods in Equine Nutrition

With diversification in the farming industry and the growth in leisure activities and tourism, there is an increasing interest in equine science, especially at colleges and universities. Whereas the ruminant has been the main focus of attention in the past, research into the relative efficiency of various equine feedstuffs is gathering momentum. The analytical chemist is usually involved with the animal nutritionist in selecting the elements or feed fractions of interest, and the levels of these substances in the feeds are important in determining or modifying the method to be used. An informative book is that by Frape (1986, pp. 35, 121, 209 and 238), which includes tables of acceptable concentrations in the feed of minerals, trace elements, crude protein and vitamins, and also the nutritional composition of four tropical grasses and Newmarket grass; blood electrolyte concentrations are also listed.

Toxic Effects of Some Elements

Lead is tolerated at 1–5 mg kg^{-1} diet, but a sustained level of about 12 mg kg^{-1} would be lethal. Lead shot trapped in silage may partially dissolve in fermentation acids to give up to 3800 mg soluble Pb kg^{-1} DM. The daisy (*Bellis perennis* L.) can accumulate 60–80 mg Cd kg^{-1} from contaminated soils, which is 30 times more than grass, therefore herbage growing in the vicinity of derelict mine workings should be analysed for heavy metals. A 500-kg horse grazing rough terrain may ingest 1–2 kg of soil per day, so the amount of any industrial fallout, contamination or seepage should be determined. Herbage high in molybdenum, iron or sulphur may lower copper absorption and thus depress serum copper. Feed is analysed by the usual procedures already described for mineral, trace and toxic elements.

Fibre

Frape observes that a horse digests fibre less easily than domesticated ruminants, therefore shorter grass containing a higher proportion of leaf is a more valuable feed than herbage approaching maturity. A suitable fibre method for non-ruminants is the NDF procedure (see Method 7.5. Determination of neutral detergent fibre (NDF) or plant cell-wall constituents, page 133).

Silage and haylage

Until recently (Moore-Colyer and Longland, 2000), little has been published regarding the feeding of silage and haylage to horses, but haylage is becoming more available commercially as farms seek to diversify. The main problem with silage is the risk of botulism (from *Clostridium botulinum*) and enteritis (from *Cl. perfringens*). Clostridial spores affect horses more than ruminants. Silage should be well fermented, high in DM and free from moulding. The problem with hay is dust which causes respiratory problems, so it is sometimes washed before feeding. Haylage lacks the dust, and is safer than silage because it is much higher in DM (approximately 50% DM), although more expensive. It is readily digested by ponies and offers a suitable high-energy alternative to hay in horse rations (Moore-Colyer and Longland, 2000). Before feeding it should be checked that the haylage or silage smells sweet (or analysed for VFAs, ensuring low values for butyric and valeric acids), and the DM and pH (which should be 4.0–4.5) determined (Frape, 1986, pp. 234–235).

Details of the DM, pH, lactic acid and VFA content in haylage are available at:

http://www.dairybiz.com/archiv/cowtalk_41.htm

and the nutrient composition of commercial semi-wilted bagged forages can be found at:

http://www.horsehage.co.uk/equine.html

Digestibility

The determination of digestibility parameters using methods designed for ruminants are clearly not applicable to horses, which have no rumen, but possess a simple stomach and an enlarged caecum and colon where micro-organisms facilitate the digestion of cellulose. Unlike in the ruminant, the products of microbial digestion have less opportunity of being absorbed and no opportunity of being further broken down by its own digestive enzymes (McDonald *et al.*, 1969).

It is the 'apparent digestibility' rather than the 'true digestibility' that is usually determined. This is because substances in the faeces not arising directly from the food lead to an underestimation of the proportion of the intake actually absorbed by the animal, also the fractions of the faeces derived from food or endogenous origins are in most cases indistinguishable (McDonald et al., 1995). Endogenous constituents include sloughed-off cells from the gut mucosa, microbial matter and digestive enzymes. Thus, for the case of crude protein:

$$\text{apparent digestion coefficient} = 100 \times \frac{(\text{N intake} - \text{total faecal N})}{\text{N intake}}$$

$$\text{true digestion coefficient} = 100 \times \frac{(\text{N intake} - [\text{total faecal N} - \text{metabolic N}])}{\text{N intake}}$$

The total faecal N is the undigested feed N in the faeces. The true digestion coefficient is not the same as the true protein, because the feed can contain nitrogenous substances, other than crude protein, which can be converted to ammonia in the Kjeldahl digestion. The metabolic N may be estimated by analysing the faeces for non-feedstuff-derived N after feeding a diet containing no protein, or only a small amount of highly digestible protein. Faeces markers have been recommended for this procedure in the case of non-ruminants (Schneider and Flatt, 1975). The addition of a solution of 3% (v/v) sulphuric acid (approximately 0.5 M) to the faeces will help prevent loss of ammonia on drying. In the case of calcium, the equations become:

$$\text{apparent digestion coefficient of Ca} = 100 - \left[\frac{\text{faecal Ca}}{\text{Ca intake}} \times 100 \right]$$

$$\text{true digestion coefficient of Ca} = 100 \times \left[\frac{\text{Ca intake} - \text{faecal Ca} + \text{endogenous faecal Ca}}{\text{Ca intake}} \right]$$

The inability to correct for the loss of gases such as methane would lead to an overestimate of food absorption, digestible carbohydrate and digestible energy, but various mathematical modelling techniques have attempted such a correction (Cone, 1997). Earlier work with ruminants by Blaxter and Clapperton (1965) gave an equation linking digestible energy with methane production based on more than 2500 determinations of the 24-h production of methane by sheep and cattle.

Methods to estimate digestibility which avoid the necessity for fistulation include gas production from feeds incubated with faecal inoculums using an automatic pressure evaluation system, and various digestion marker methods (see below). The concentration of marker in the feed and faeces may be used to calculate the apparent digestibility of nutrient components (such as crude protein) by employing the following equation:

$$\% \text{ nutrient apparent digestibility} = 100 - \left[100 \times \frac{\% \text{ marker in feed}}{\% \text{ marker in faeces}} \times \frac{\% \text{ nutrient in faeces}}{\% \text{ nutrient in feed}} \right]$$

Digestion markers

The rate of passage through the gastrointestinal tract exerts an important influence on the intake and digestibility (Poppi et al., 2000). Thus, digestibility is reduced as the rate of passage increases (Pearson and Merritt, 1991). Finely grinding the feed decreases the retention time and hence the digestibility (Blaxter *et al.*, 1956). Mean retention times may be monitored by administering a chemical marker with the feed and collecting the faeces over a defined period; this avoids the difficulty of total collection procedures. Naturally occurring largely indigestible substances, such as lignin, acid-insoluble ash and indigestible ADF have been tried as markers, but are not very satisfactory. A marker should remain unchanged after passing through the digestive tract, neither should it migrate. Migration is when the marker becomes detached from the food particle and then the unbound marker reattaches to other food particles. The marker should ideally not be attached to the whole continuum of particle sizes, because there may be selective retention of larger or smaller sized particles in some sections of the gut (Cork *et al.*, 1999). It is essential to recover the metal marker from the same food or plant matrix to which it was initially bonded. Chromium (III) oxide (chromic oxide, Cr_2O_3) powder may be administered in a capsule, but is no longer considered suitable, because it does not associate specifically with either the particulate or liquid component of the ingesta and results in sedimentation and sporadic transfer of marker (Bertone *et al.*, 1989a). The soluble chromium (III) salts bond only slowly to the feed surface, and therefore it is preferred to reduce a chromium (VI) salt (e.g. dichromate) with ascorbic acid to bind it on to the fibre by a mordanting process as explained by Marais (2000). Cobalt and titanium have been used, but current opinion, however, favours the use of rare earth elements, cerium, dysprosium, and ytterbium, of which ytterbium gives the most favourable results. The ytterbium solution is best applied by soaking, rather than spraying, the feed with a solution of the ytterbium salt, because this exposes the ytterbium ions to more of the natural adsorption sites (Mader *et al.*, 1984). The salt which has been used for horses is ytterbium (III) chloride hexahydrate (Aldrich, mol. wt 387.54) (Bertone *et al.*, 1989b). Subsequent analysis may be by AAS (Teeter *et al.*, 1979; Bertone *et al.*, 1989a; Luginbuhl *et al.*, 1994) or ICP-MS (Combs and Satter, 1992). Ytterbium as a marker has been used experimentally and reviewed in a thesis by Morrow (1998), where big-bale silage and hay were fed to ponies, and the rate of passage determined. Although little significant difference existed between the feeds, the trend was for the silage to have longer mean retention times than the hays.

Method 10.1a. Application of ytterbium marker to feed

Reagent.

- Ytterbium marker solution – dissolve 10 g ytterbium (III) chloride hexahydrate (Aldrich Chemical Co., mol. wt 387.54) in 1 l distilled water and adjust the pH to 3.8 using 0.1 M HCl.

Procedure. Weigh 100 g fresh weight of the feed material, previously chopped or unchopped as required for the experiment, and suspend in 1 l of the ytterbium marker solution for 24 h. Subsequently rinse the feed material in distilled water and re-soak in fresh distilled water every hour for 6 h to remove all unbound ytterbium ions. Pour off the water and dry at 50°C until the weight is approximately 100 g, i.e. the same as the original weight of fresh feed material. Accurately weigh approximately 15 g subsample and dry at 60°C to constant weight, weigh and calculate the dry matter. Mill the dried marked feed to 1 mm and retain for ashing before Yb analysis.

Values for the resulting ytterbium concentration in DM for some feeds and corresponding faeces are given by Morrow (1998):

Big bale silage (long chop-length)	17,637 µg g^{-1}
Big bale silage (short chop-length)	29,178 µg g^{-1}
Hay (long chop-length)	12,922 µg g^{-1}
Hay (short chop-length)	18,052 µg g^{-1}

As would be expected, the smaller the feed particle size, the greater the surface area per given weight is available for bonding to the ytterbium ions, hence the greater the concentration of marker. The greatest concentration of ytterbium in faecal DM occurred after approximately 24 h, with the concentration approximately zero after 75 h; the mean retention time was 26.07 h. The maximum ranges for the concentration of ytterbium in the faecal DM after approximately 24 h was approximately as follows:

Big bale silage (long chop-length)	500–800 µg g^{-1}
Big bale silage (short chop-length)	300–500 µg g^{-1}
Hay (long chop-length)	500–750 µg g^{-1}
Hay (short chop-length)	750–850 µg g^{-1}

Calculation of mean retention time. The mean retention time (MRT) may be calculated using equations given in Blaxter *et al.* (1956) or the simplified version in Pearson and Merritt (1991):

$$MRT = \Sigma M_i t_i / \Sigma M_i$$

where M_i is the concentration of marker excreted in the faeces at time t_i following the administration of the marked feed. This equation was also used for studying flow rates in equines by Nyberg *et al.* (1995).

Method 10.1b. Feeding of ytterbium marked feed and faecal collection and preparation

Procedure. Depending on the diet allocation, weigh an appropriate quantity of the remainder of the marked feed and feed to the specified animal before its normal feed, ensuring all the marked feed has first been consumed. This could be at 21.30 h on the Monday of the collection week. Collect faeces samples 1.5 h before the first feed, then after 8 h, 12 h, and 18 h, and then at 2-h intervals until 01.30 h on Thursday, and at approximately 4-h intervals until 21.30 h Friday. A final sample is taken at 11.30 h on Saturday.

Collect the faeces from the stable floor at the above times, place in a tared bucket and calculate the total weight of faeces. Thoroughly mix and transfer an approximately 200 g subsample to a labelled grip-top polythene bag, and place in a deep freeze. The rest is transferred to the dustbin allocated to the particular animal. Remove the sample from the deep-freeze and allow to partially thaw. Take a 10 g aliquot for determination of DM by drying to constant weight at 60°C. This is subsequently milled to 1 mm and transferred to a sample tube before preparation for Yb analysis. Take a further 100 g portion and freeze dry, then mill to 1 mm and store in an airtight container. Weigh a 10 g sample for determination of DM in the freeze-dried material at the same time as weighing aliquots for determination of nutrient components, e.g. ADF, NDF and CP (crude protein) as required. The bulked faeces may also be similarly prepared for analysis.

Method 10.1c. Preparation of ytterbium marked feed for analysis

Reagent.

- Nitric acid solution, approximately 5% (v/v) – add 100 ml nitric acid (approximately 70% m/m HNO_3) to 2 l distilled water and mix.

Procedure. Weigh 2 g of the oven dried sample ground to 1 mm into a porcelain or glass crucible and place in a cool muffle furnace. Increase the temperature to 550°C and ash for 2 h. Cool and dissolve in nitric acid solution, make up to 20 ml and store in a polythene capped tube. Allow to settle before transferring to another clean sample tube by decanting or using a polythene Pasteur pipette. Analyse the solution by AAS or ICP according to the manufacturer's recommendations.

Note: A typical ICP analysis requires making a solution of 9.6 ml ultra-pure water, 0.2 ml nitric acid (approximately 70% m/m HNO_3), 0.1 ml marked feed sample solution (or 1 ml faecal sample solution) and 0.1 ml rhodium internal standard solution (10 µg ml^{-1} in 5% v/v nitric acid solution). This is contained in a sample tube sealed with clingfilm. The sample probe is inserted through the clingfilm. A pick-up time of 35 s, and a scan time of 35 s proved satisfactory. The probe is rinsed in the 5% v/v nitric acid solution between samples.

Mobile Bag Technique (MBT)

Some studies use the mobile bag technique in fistulated animals, but no fistulation is necessary when the small bags of feed are discharged into the stomach via a nasogastral (naso-oesophageal) tube. It was first used in equids by Machelboeuf *et al.* (1995), and subsequently by Hyslop and Cuddeford (1996). An assessment of this technique for studying the dynamics of fibre digestion in equids has been made by Tomlinson (1997). She used rye grass hay, dehydrated grass, dehydrated lucerne and grass chaff, and concluded that the MBT is a useful tool for studying the dynamics of total tract digestion in equids, and is sensitive enough to distinguish differences in digestibility between the four foodstuffs. A typical graph of disappearance of DM and NDF from mobile bags containing ryegrass hay subjected to total tract digestion by a mature Welsh cross pony gelding is shown in Fig. 10.1.

The MBT is one of the *in sacco* methods, and these have been reviewed by Nozière and Michalet-Doreau (2000). The bags are made from polyester, nylon or Dacron material woven with a controlled pore size and heat-sealed into small pouches. The latter is a compromise between retention of the ground foodstuff and sufficient permeability to allow complete penetration of the food particles by the gastric juices, enzymes and microbial population. A range of between 40 μm and 60 μm is suitable for most purposes (Nocek, 1988). The sample size should be ≤1.0 g, and milled to 1 mm. Machelboeuf *et al.* (1995, 1996), however, used 200 mg milled to 3 mm, and sealed them in 60 × 10 mm bags. Cherian *et al.* (1988, 1989), found that a pre-digestion stage gave closer agreement with conventional methods when measuring the apparent digestibility of protein for swine. This was achieved by soaking the feed, with agitation, for 4 h in a solution of pepsin with an activity of 3771 IU l^{-1} in 0.01 M HCl (pH 2). The exposure of samples to equine saliva before insertion into the stomach may affect the digestibility, but is yet to be investigated. There is, however, little amylase in equine saliva (M. Moore-Colyer, Aberystwyth, 2001, personal communication).

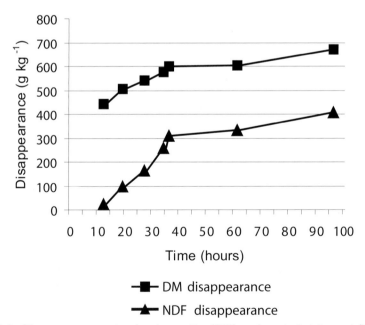

Fig. 10.1. Disappearance rates for dry matter (DM) and neutral detergent fibre (NDF) from ryegrass hay samples in polyester mobile bags subjected to total tract digestion in a pony (adapted from Tomlinson, 1997).

Transit time

A number of bags are discharged into the stomach simultaneously, yet one needs to have a range of transit times to be able to plot transit time versus quantity of nutrient solubilized in studies of digestion dynamics. The best technique to achieve this is the use of different size bags, where the larger bag slows the transit time. When comparing digestibilities of feeds, however, the transit times should ideally be identical, therefore any variation must be corrected for.

Method 10.2. Determination of digestibility using the mobile bag technique

Equipment.

- Monofilamentous polyester fibre – 0.41 µm pore size, (Sericol Ltd, Westwood Road, Broadstairs, Kent CT10 2PA, UK), folded in half and double-sealed along two of the three open sides using a fabric heat-sealer to make bag sizes of 40 × 10 mm and 60 × 10 mm.
- Nasogastral stomach tube – approximately 10 mm ID × 2500 mm long flexible PVC tubing with rounded end for insertion.

Note: There are several Home Office regulations under the Animals (Scientific Procedures) Act 1986 which must be complied with in the UK. The premises used for the research project will require a certificate of designation to authorize its use for the specified animal species and procedure. A project licence must be obtained for the specific procedure/s being undertaken; details must be given regarding the background, objectives, benefits, plan of work, list of procedures, severity of impact on the animal, and whether any pain, suffering, distress or lasting harm may occur. Finally, a personal licence will need to be obtained for the person performing the procedure, and details provided of the techniques, animals, and whether anaesthesia will be administered. It is also wise to take veterinary advice before proceeding. For further information and application forms, consult the Home Office website pages:

> http://www.homeoffice.gov.uk/animact/aspileaf.htm
> http://www.homeoffice.gov.uk/animact/animform.htm

Procedure. Grind approximately 100 g foodstuff through a 1-mm mesh and sieve to remove all particles (about 0.5–2.0% by weight) less than 45 µm. Number the bags with a waterproof pen, dry the bags at 60°C for 8 h, cool in a desiccator and weigh. Transfer 130 mg foodstuff into the 40 × 10 mm bags and 200 mg into the 60 × 10 mm bags, double-seal the open end and reweigh to confirm no sample has been lost. Also weigh a separate sample of foodstuff for DM determination by heating at 60°C for 48 h, and a sample for NDF, or other nutrient components, if required. If weighing the same foodstuff, take a further portion for DM after every batch of 22 bags. Samples should also be taken and subjected to only the washing cycle to give a zero time value and an estimate of water-soluble components plus loss of any fine particles. Load the stomach tube by inserting the batch of bags with a brass rod with rounded ends or similar, being careful not to puncture the bags. Using a hand pump, flush the sample bags into the stomach using approximately 750 ml warm water during the morning meal after ingestion of approximately 1 kg hay, then continue the meal.

Retrieve the bags from the faeces at the set times; if there is any delay between retrieval of bags and washing, or between washing and drying, store the bags at 4°C to reduce microbial activity. Briefly rinse under running tap water to clean the bulk of the material adhering to the outside of the bags, then wash the bags in cold water in an automatic washing machine (Hyslop *et al.*, 1999) for 45 min with four rinses (4 min immersion in water plus 3.3 minutes agitation per rinse) to remove the remaining adhering microbial and exogenous debris from the outer surface, also endogenous and free microbial contamination from within the bag. (*Note*: hand-rinsing in tap water may lead to significant variations, and ⩾5 short term rinses, e.g. 1 min agitation plus 2 min spin have been suggested (Broderick and Cochran, 2000); machine washing undigested samples in bags reveal there could be a significant DM loss from this process, but this loss would include water-soluble components. A less severe washing methodology may need to be developed.) Dry the bags in a forced-draught oven at 60°C for 48 h, cool in a desiccator and weigh to determine the disappearance in DM. Remove the residue for determination of NDF etc., if required. The residual DM from bags containing the same foodstuff retrieved from a particular animal on the same occasion may be pooled and well mixed for subsequent analyses.

Data analysis. The data may be analysed using the techniques presented by Ørskov and McDonald (1979) or Dhanoa (1988). The former authors suggested fitting the rate of disappearance (*p*) of DM, CP, NDF, etc., to an equation of the form:

$$p = a + b(1 - e^{-ct})$$

where *t* is the incubation time. With increase in *t*, *p* increases, but at a reducing rate. This is an empirical equation to fit the incubation data, where *a*, *b*, and *c* are constants fitted by an iterative least-squares procedure. It is also possible to conceive that these constants represent the following parameters:

> *a* is the rapidly soluble fraction
> *b* is the slowly degradable fraction
> *c* is the fractional rate constant at which *b* will be degraded per unit time

For a protein supplemented feed, Ørskov and McDonald (1979) found $a = 20$, $b = 80$, $c = 0.082$ and $k = 0.046 \text{ h}^{-1}$.

The effective degradability, or effective percentage degradation (ED), becomes:

$$ED = a + [bc/(c + k)]$$

where *k* represents the fractional outflow rate passing through the gut per hour, which is measured by regression analysis. In ruminant studies, these authors determined *k* in a separate experiment using a chromium marker to render indigestible the feed supplement particle to which it was attached. ED therefore provides an estimate of the degradability of the feed component under the specified feeding conditions.

Hemicellulose

This is unfortunately a misnomer, as the cell-wall hemicellulose fraction has no chemical relationship with cellulose. In the primary and secondary cell-walls, the cellulose microfibrils are embedded in an amorphous matrix consisting of pectins and hemicelluloses. Pectins are mainly polygalacturonic acids and predominate in the middle lamella between neighbouring cells; this is also the most highly lignified plant tissue (Coleman *et al.*, 1999). Hemicelluloses are short chain partially soluble polysaccharides composed of a variety of sugars such as arabinose, galactose, glucose, mannose and xylose in combination with uronic type acids, such as glucuronic and galacturonic acid. Hemicelluloses that are basically xylose or arabinose in combination with glucuronic acid are termed xyloglucans or arabinoglucans, respectively. If arabinose and galactose are in approximately equal amounts, the hemi-cellulose is termed an arabinogalactan; arabinose and xylose combine to give arabinoxylan, which has few side chains. If mainly just glucose, galac-tose or xylose, they are called glucans, galactans or xylans respectively. Hemicellulose molecules, such as galactoarabinoxylan, are often branched and, like pectic compounds, very hydrophilic; they become highly hydrated and form gels. The hemicellulose in *Poaceae* (= *Graminae*, grasses, cereals) contains arabinoxylan, galactoarabinoxylan and glucan. Hemicellulose is abundant in primary walls but is also found in secondary walls.

The estimation of total hemicellulose, without characterization of indi-vidual sugars, may be achieved by subtracting the ADF value from the NDF value obtained from duplicate samples of the same weight. There are, how-ever, four main possible interferences with this procedure.

1. Biogenic silica is largely dissolved by NDF method, but totally recovered with ADF.
2. Cell-wall protein is recovered in the NDF method, but mostly dissolves with ADF.
3. Pectin is dissolved in the NDF method, but is part precipitated with ADF.
4. Tannin is part dissolved by NDF, but precipitated as a protein complex with ADF.

It should be determined whether any of these factors is large enough to constitute a significant interference in the planned experiment. See also the comments on NDF below. An NDF method which includes pectic poly-saccharides, β-glucans and fructans has been published by Hall *et al.* (1997).

Non-starch Polysaccharides (NSP)

Plant carbohydrates may be divided into three groups as shown in Table 10.1. The non-starch polysaccharides, e.g. cellulose, fall mainly into the third group of structural carbohydrates. (Although the hemicelluloses such as galactans and mannans are associated with the cell-wall, they are termed storage polysaccharides (Graham, 1991), and have been included under the structural category in Table 10.1 for convenience.) They are structural cell-wall carbohydrates associated with the fibre fraction. Although the NDF determination is suitable for members of the *Poaceae* (= *Graminae*, grasses, cereals), it underestimates the cell-wall content in legumes. One reason is that legumes and other non-grass species contain relatively high concentrations of pectic polysaccharides that are solubilized by neutral detergent and therefore omitted

Table 10.1. The three main groups of plant carbohydrates with typical examples and occurrences.

Carbohydrate group	Sub-divisions; other descriptions	Examples	Typical occurrence
Soluble sugars	Monosaccharides	Glucose	Plants, fruit
		Fructose	Green leaves, fruit
	Oligosaccharides (2–9 monosaccharide residues)	Sucrose (disaccharide)	Sugar beet, carrots
		Maltose (disaccharide)	Germinating barley
		Raffinose (trisaccharide)	Sugarbeet, molasses, cotton seed (8%)
	Polysaccharides (soluble storage)	Fructans (fructosans)	All plant parts of *Compositae* and *Poaceae*; 50% DM in Jerusalem artichoke
Starches	Storage polysaccharides	Amylose (20–28%) (linear molecule) together with amylopectin (72–80%) (branched molecule) (Note: both are glucosans)	Seeds, tubers, roots
Structural (including cell wall storage polysaccharides)	Non-starch polysaccharides; cell-wall/fibre carbohydrates; lignocellulose	Cellulose (a glucosan)	Cell-walls, cotton
		Arabinoxylans (hemicelluloses)	Cereals
		β-glucans (glucosans, mixed-linked)	Endosperm cell-walls of barley and oats
		Galactans, mannans (hemicelluloses, cell-wall storage)	Palm kernels, lupin seed
		Pectins (partially soluble)	Dicotyledons
		Xylans (hemicelluloses) (Note: hemicelluloses and pectic substances are heteropolysaccharides)	Grasses

from the NDF fraction (Chesson, 2000); some of the hemicellulose fraction is also omitted for the same reason. For hind-gut fermentors, such as equids, the NSP method is also to be preferred, because arabinoxylans, β-glucans, oligosaccharides, and xylans are only fermented in the hind-gut. Different species, or breeds within species, may digest these individual components with varying degrees of efficiency, therefore their individual characterization and estimation would be beneficial in animal nutrition studies.

Starch that escapes amylolytic digestion in the small intestine, together with the oligosaccharides, fructans and NSP, will proceed to the large intestine for fermentation by hind-gut microflora. If present to excess, sufficient lactic acid will be produced to lower the pH in the hind-gut, possibly causing colic or laminitis. Up to about 15% of the NSP has been found to disappear pre-caecally, and is therefore unavailable for microbial breakdown and absorption of nutrients in the lumen of the hind-gut (Moore-Colyer *et al.*, 1997a, b). The determination of the different classes of carbohydrates in the feed will help in the correct formulation of horse rations (Longland, 2001, unpublished).

Method 10.3. Determination of total non-starch polysaccharides

This is based on the procedure by Englyst and Cummings (1984) who performed a GLC analysis of the alditol acetate derivatives of the constituent sugars. It incorporates slight modifications as currently carried out at IGER, Aberystwyth (Paul Thomas, IGER, Aberystwyth, 2001, personal communication). The starch is first dispersed with dimethyl sulphoxide, which disrupts intermolecular hydrogen bonds, followed by hydrolysis with α-amylase and pullulanase. The former hydrolyses the straight chain α-1–4-glycosidic bonds of amylose, the latter, also known as alphadextrin 6-glucanohydrolase, is a specific enzyme for the hydrolysis of the branching α-1–6-glycosidic bonds in the amylopectin component of starch.

Reagents.

- Acetate buffer – 0.1 M sodium acetate solution adjusted to pH 5.2.
- Acetone
- β-D-Allose internal standard solution, 1 mg ml^{-1} – dissolve 200 mg β-D-Allose (Sigma Cat. No. A-6390) in 50% (v/v) saturated benzoic acid solution and make up to 200 ml. Store in the dark. (This is sufficient for at least 36 tests – it is expensive, so make just a sufficient amount.)
- Dimethylsulphoxide – (DMSO, Sigma ACS Reagent, Cat. No. D8779).
- Ethanol (absolute)
- Ethyl acetate
- Mixed enzyme solution – either make up a mixed enzyme solution containing 5000 units of α-amylase and 5 units of pullalanase per millilitre acetate buffer (original method), or, as is usual, separately add 0.5 ml α-amylase reagent and 0.1 ml pullulanase reagent per sample.

- α-Amylase reagent – (EC 3.2.1.1.), capsules (Pancrex V, 300 capsules, high potency pancreatin. Contains: amylase, 9000 BP units; lipase, 8300 BP units, free protease, 430 BP units. Manufacturer: Paines and Byrne Ltd, West Byfleet, Surrey, UK; Supplied by Boots Chemists). For 32 samples, immediately before use dissolve four capsules in 18 ml water at room temperature and centrifuge.
- Pullulanase reagent – (EC 3.2.1.41.), suspension in 3.2 M $(NH_4)_2SO_4$, pH 6.2, 5 units per mg protein (Sigma Cat. No. P5420). For up to ten samples, make up 0.010 ml pullulanase suspension to 1 ml with acetate buffer, pH 5.2, and keep refrigerated until use.
- Sulphuric acid, 12 M H_2SO_4

Equipment.

- Three-place magnetic stirrer-hotplate
- Vortex mixer

Procedure (enzyme and acid hydrolysis). Weigh accurately 50–1000 mg representative sample (freeze dry, ball mill or homogenize if necessary, but do not oven-dry) containing ≤150 mg starch and ≤50 mg NSP into a 50–60-ml screw-top centrifuge tube (borosilicate glass, approximately 200 × 26 mm) and add an approximately 12 mm PTFE-coated magnetic stirring bar. Samples with 90–100% DM and <2–3% fat may be analysed directly, which is usually the case, otherwise add 40 ml acetone, stir for 30 min, centrifuge and remove as much supernatant as possible, without disturbing the residue, by means of a plastic Pasteur pipette or glass capillary connected via a Buchner flask (to retain the solvent) to a vacuum line. Evaporate off the acetone using a water-bath at approximately 65°C in a fume cupboard. (This may be an arrangement of three 2-l glass beakers of water on a three-place magnetic stirrer-hotplate, which will allow mixing until dry. A J-cloth may be placed in the bottom of the beaker to prevent breakages, and the beaker should be covered with a watchglass to reduce evaporation when used as a boiling water bath (see below). Occasionally check the water level to ensure it does not boil dry.)

Add 2 ml DMSO to the centrifuge tube, replace the cap and heat for 1 h after returning to the boil in a boiling water bath on a stirrer-hotplate (as described above) with continuous stirring. Without cooling, from a dispenser add 8 ml acetate buffer at 50°C and vortex mix immediately. After cooling to 45°C, immediately add 0.5 ml α-amylase reagent and 0.1 ml pullulanase reagent. Incubate for 16–18 h at 45°C with regular mixing. Remove from the incubator and add 40 ml absolute ethanol, mix thoroughly and allow to stand for 1 h at ambient temperature before centrifuging for approximately 10 min, or until a clear supernatant is obtained. Remove as much of the supernatant as possible by aspiration into a Buchner flask for safe disposal. Wash twice with 50 ml 85% (v/v) ethanol by mixing well, centrifuging, and removing the solvent as above. Add 40 ml acetone to the residue, and stir thoroughly on the magnetic stirrer-hotplate for 5 min. Check that all the stirrer-bars are rotating or oscillating well, and if necessary, adjust the position of the beaker to achieve this. Aspirate the solvent into a separate Buchner flask for safe disposal or redistillation and re-use. Dry in a beaker of water at 65°C in a fume cupboard.

Add 2 ml of 12 M sulphuric acid to the dried residue and disperse using a vortex mixer; inspect the bottom of the tube to check that no unmixed sample remains. Heat at 35°C for 1.25 h to solubilize the cellulose, then quickly dispense 22 ml water and mix. Heat in a boiling water bath for 2–2.25 h timed from re-boiling, with continuous magnetic stirring. Place in water at room temperature to cool, and add 5 ml β-D-allose internal standard solution and mix; this solution may be refrigerated until required for analysis. Take a 1-ml aliquot of the hydrolysate for preparation of alditol acetates, and retain the rest if required for optional determination of interference from uronic acids (see Englyst and Cummings, 1984).

Preparation of alditol acetates

Reagents.

- Acetic anhydride
- Ammonia solution, 12 M and 3 M – dilute 92 ml of ammonia solution, 0.910 g cm³, approximately 13 M, to 100 ml with water to give a 12 M solution. Further dilute 25 ml to 100 ml to give a 3 M ammonia solution.
- 1-methylimidazole – (N-methylimidazole, NmetIm; Sigma Cat. No. M8878, 500 ml) *Safety note*: extremely dangerous to mucous membranes of upper respiratory tract, burns mouth, oesophagus and skin, and destructive to eyes. Use only in a fume cupboard/hood. Flush contaminated skin with copious amounts of water.
- Mixed standard – dissolve 200 mg arabinose, 100 mg galactose, 700 mg glucose, 100 mg mannose, 100 mg rhamnose and 500 mg xylose in 50% saturated benzoic acid solution and make up to 100 ml; keep refrigerated in the dark – it has a long shelf-life.
- Octan-2-ol
- Potassium hydroxide solution, 7.5 M – dissolve 210 g potassium hydroxide pellets in water and make up to 500 ml and mix. *Safety note*: this is highly caustic and PPE must be worn.
- Sodium tetrahydroborate(III) reagent – freshly prepare by dissolving 50 mg sodium tetrahydroborate(III) per ml 3 M ammonia solution (0.1 ml reagent used per sample). *Note*: original method used 100 mg ml^{-1} concentration.

Equipment.

- Crimp-top vials – suitable vials for GLC are available in the UK from various suppliers, including Vials Direct, Cat. No. V2.3T, PO Box 117, Macclesfield SK11 8DH.

Procedure. Transfer 1 ml hydrolysate solution to an approximately 100 × 21 mm borosilicate centrifuge tube and add 0.2 ml of 12 M ammonia solution and mix; test a small drop to ensure that it is alkaline. Add 0.1 ml of a freshly prepared solution of sodium tetrahydroborate(III) reagent and 1–5 μl of octan-2-ol to prevent foaming, then vortex well to mix. Heat for 1 h at 40°C. Next add 0.1 ml glacial acetic acid, vortex-mix and remove a drop to check the pH is acid. Transfer 0.5 ml of the acidified solution to a smaller screw-cap glass tube and add 0.5 ml 1-methylimidazole, 5 ml acetic anhydride, vortex-mix and leave at room temperature for 10 min. Add 0.9 ml ethanol, vortex-mix, leave 5 min before adding 5 ml water and then vortex-mix. Place in an ice-water bath for 5 min, then add 5 ml 7.5 M KOH solution, leave 5 min, then add a further 5 ml 7.5 M KOH solution. Cap the tube and invert to mix, then leave the tube to allow the two liquid phases to separate. Transfer the top solvent layer to a small vial, crimp on the top and store at 5°C before injecting about 1–2 μl onto the gas chromatograph.

Take 0.25 ml of the mixed sugars standard solution, add 0.25 ml allose internal standard solution and 0.5 ml 2 M sulphuric acid, mix and substitute instead of the hydrolysate solution in the above procedure for preparation of alditol acetates. The concentration of sugars should give a linear relationship with peak area over the normal ranges. The GLC determination conditions will depend on the instrument, but the following are suggested:

Injector temperature:	260°C
Oven temperature:	220°C
Detector (FID) temperature:	250°C
Carrier gas:	helium.

11 Methods for Organic Farmers and Growers

There are two main questions people have asked about this chapter: 'Why should soil analyses for organic farming differ from those for conventional farming?' and 'How can you get the details of methods and interpretation of results when little has been published, and there is so much secrecy because of vested commercial interests?' The former query is easier to tackle than the latter. It will be seen that the organic approach to soil chemical analysis looks at nutrient balance and potential availability over a longer term, rather than the immediate availability of nutrients. The commercial laboratory of Dr Friedrich M. Balzer has kindly provided some details of his designated methods, with corresponding guidelines on interpretation of results indicated on their website. More detailed information from commercial laboratories on interpretation, particularly from those using the Albrecht approach, would be desirable.

Origins

Before the availability of artificial fertilizers in the mid-19th century, farms were traditionally organic, with recycling of animal waste, and perhaps with the application of lime on acid soils. Agricultural chemical analysis may have begun with Carl Wilhelm Scheele (1742–1786), the Swedish pharmacist who isolated citric acid from lemons and gooseberries and malic acid from apples. In France, Nicolas Theodore de Saussure (1767–1845) studied the mineral composition of plant ash, and in Britain, Sir Humphrey Davy

(1778–1829) analysed plants into 19 constituents, estimated the feeding value of 97 different grasses, and published his *Elements of Agricultural Chemistry* in 1813 (Faithfull, 1993).

The term 'Father of Agricultural Chemistry' is usually ascribed to Baron Justus von Liebig (1803–1873). In 1840, Liebig withdrew from the sphere of pure organic chemistry to apply his genius to the study of agricultural chemistry. He finally put the nail in the coffin for the humus theory, which claimed that plants mainly derived their carbon from humus in the soil rather than carbon dioxide. The role of nitrogen-fixing bacteria on the nodules of leguminous plants was only presented in 1886 and published in 1887, after Liebig's death. He therefore was under the misapprehension that most of the plant's nitrogen supply originated in ammonia from the air. He was aware, however, of the efficacy of legumes as nitrogen gatherers (Curtis, 1942). Another deficiency was the fact that he never took the acidity of the soil into account (Bradfield, 1942). Liebig assiduously carried out analyses of the mineral components of plant ash from the viewpoint that if the mineral elements removed from the soil by a particular plant species could be replaced or increased by application of a fertilizer compounded in the same proportions of mineral elements as found in the ash, then the yield would be enhanced. He considered the minerals to be in the form of solutions held in a state of physical absorption within the soil, the role of the cation exchange properties of clay and humus in the soil not yet being discovered. In this context, (Walters, 1989, p. 161), Albrecht quotes S.C. Hood who said, 'The oldest and most persistent of these errors may be referred to as the Liebig Complex. Over 100 years ago Justus von Liebig announced that plants needed from the soil no more than proper amounts of nitrogen, phosphorus and potash in water-soluble forms' (Liebig, 1840).

Balance

Liebig's important contribution, in spite of the above misconceptions, was to express the earlier concept of mineral balance proposed by Sprengel (1787–1859), which he formulated as the Law of the Minimum (Browne, 1942). This states that if one of the plant nutrients is present in the soil in a state of deficiency or unavailability, it will render the other nutrients inactive or lessen their activity. There are limitations to this rule, and later experiments showed that some plants will still grow using an incomplete fertilizer, but the yield will be diminished and the crop contain a deficiency of the omitted element. Other growth factors such as rainfall, sunshine and temperature should also be taken into account when defining conditions to achieve the optimum yield. Balance was later expressed as the general principle of multi-causation in nature, and more recently as that of taking a holistic view of the soil–plant ecosystem, including the soil microbial population.

Albrecht

An important figure in popularizing agricultural chemical analysis, especially regarding soils and their role in plant, animal and human nutrition and health was William A. Albrecht (d. 1974), who published hundreds of papers and press articles from 1918 to 1974. He was based at the Department of Soils of the Missouri Experiment Station. A collection of many of these publications has been edited by Walters (1975–1996). They are published by Acres USA:

> http://www.acresusa.com/original/

A brief summary of the contents is viewable at:

> http://www.metrofarm.com/store/bkreltech_1.html

and Vol. 1, covering the main concepts, is described by, and available from amazon.com.

Basic cation saturation ratio

From the early 1930s, his research often involved the study of the colloidal clay fraction, its cation exchange properties, and the optimum percentages (or ratios) of basic cations for balanced plant nutrition. This is now called the basic cation saturation ratio (BCSR) approach. He emphasized the fact that although the adsorbed cations were insoluble in water, and therefore resistant to being lost by leaching, they were nevertheless available to the plant. His values for CEC ranged from 10–80 for clays to 100–200 for organic matter. Albrecht emphasized the need for a balanced soil fertility to promote healthy plants, and found it helped prevent fungus rot in onions and nematode attack in carrots (Walters, 1989).

Many of Albrecht's findings were related to soybeans, so care must be taken in their application to other species. He derived, however, a general set of ratios of basic cations and the hydrogen ion as percentages of the cation exchange capacity which would give a balanced plant nutrition: calcium, 60–75%; magnesium, 10–20% (7–15% in some plants); potassium, 2–5%; sodium, 0.5–5%; hydrogen, 10% (>10% is an acid soil); other cations (essential trace elements), 5%. He regarded these figures as guidelines applicable to humid region soil treatments for legumes, but a sound reasoning basis for the better growth of non-legumes (Albrecht, 1967). Walters (1996) says that the nutrient code for cations expressed above is being used by the important laboratories serving eco-agriculture.

Albrecht campaigned against the concept of an acid soil causing poorer crop growth; rather, it is the calcium deficiency that needs to be remedied. The acid soil solution dissolves rock particles, such as rock phosphate and limestone, to release beneficial nutrients such as phosphate and calcium respectively. It also mobilizes the other adsorbed ions off the clay–humus colloid. He estimated the optimum ratios of calcium to magnesium and calcium to potassium. These were approximately from 4:1 to 7.5:1, and from 15:1 to 38:1 respectively. The higher the Ca:K ratio, the more proteinaceous the soybean crop; the lower the ratio, the more carbonaceous, with a higher yield, but lower P and Ca. He would not recommend any other basic cation ratios, and added that the ratios should be adjusted by fertility treatments for the most efficient plant nutrition (Walters, 1989). Excesses of individual cations should be avoided to prevent harmful effects: thus excess Ca or K reduces the transport of Mg into the crop; excess Mg reduces K in the crop, and excess Ca reduces the uptake of B, Fe, K, Mn and Zn. Albrecht helped E.R. Kuck design a soil audit and inventory report for the Brookside Laboratories of Brookside Dairy Farms, New Knoxville, Ohio, USA; a facsimile is given in Walters (1996). These laboratories are still offering an analytical service and may be visited at:

> http://www.blinc.com/bli/agricult/index.html

Other ratios

C:N (or carbohydrate:protein) ratios are important with respect to the relative requirements of plants and microbes. Straw has a C:N value of approximately 80:1, whereas after ploughing under to form a humus-rich soil, the value narrows to about 12:1. Albrecht found that the humus fraction of the experimental Sanborn Field, cultivated over 50 years, had a C:N ratio of about from 2:1 to 3.4:1, where the lower ratio is similar to that of the microbes themselves. Therefore, if microbes feed on straw that has been incorporated into the soil, they will require an additional source of nitrogen, and happen to be more successful at competing with plants for the same nutrient. In Missouri clay, Albrecht found 1.5% C and 0.15% N, which represents a favourable ratio of 10:1, and is an average value for well-weathered soils (Walters, 1989).

N:K balance

It is recognized that crop response to applied N is below the optimum unless there is a sufficient supply of potash. Although specific ratios are not given, there is helpful information in free leaflets, including PDA (1999a, b) and PDA (2001), obtainable from the Potash Development Association:

> http://www.pda.org.uk

Trace elements

Albrecht recognized that trace elements are significant for plants, but are present 'in such small amounts that our measures yet designed are too unreliable to warrant specifications of them for either soil or crop' (Albrecht, 1963, and in Walters, 1989). Analytical techniques have since improved sufficiently to allow such measurements, but we are unable to find recommended amounts or ratios in Albrecht's research. He did find, however, that a BCSR of 2% K was the limit for uptake by plants unless B was present at a level of

1 mg kg⁻¹, when the limit increased to a maximum of 20% K (Albrecht, 1960, and in Walters, 1989). A lack of boron prevented the formation of nodules on lucerne, but its presence increased the content of tryptophane and lysine.

Fertilizers

Although Albrecht has been endeared to advocates of organic farming methods for his balanced holistic view, he was pragmatic when it came to the use of artificial fertilizers in addition to manures. Thus he writes, 'Fertilizer use should not serve to divert attention from manure conservation, its maximum production, and its wisest use. All possible practices in better soil management should be exercised first and then fertilizers purchased and added to make up the deficiencies in soil fertility that need to be balanced for most effective crop production. Manure use represents putting back much of what came from the soil. Fertilizer use represents putting on some fertility purchased and brought from outside the farm, to add to the soil's supply' (Albrecht, 1942, and in Walters, 1992).

Current organic farming methods seek to reduce bought-in manures to a minimum, but some fertilizers are allowed as permitted inputs on a restricted basis (Lampkin and Measures, 1999). In general, only fertilizers that release nutrients through an intermediate process, such as chemical weathering or the activity of soil organisms are allowed (Stockdale *et al.*, 2000).

Commercial Analytical Services

It is difficult to determine the validity of the wide range of methods and interpretations, claimed to be based on the Albrecht system, that are provided by commercial laboratories while they remain unpublished. Albrecht worked with US soils and a limited range of crops, and agronomic advice can only be effective if backed up by field trials under conditions prevailing in one's own country. Loveland admits that one area where the BCSR method is undoubtedly right is in its advocacy of the use of organic matter in crop rotations, even if the benefits are hard to quantify (P.J. Loveland, Cranfield University, Silsoe, 2000, personal communication. See http://www.silsoe. cranfield.ac.uk). For views favouring the Albrecht-BCSR system and details of analytical services in the USA based on this method, see the soil management page of the Appropriate Technology Transfer for Rural Areas website at:

http://www.attra.org/attra-pub/soilmgt.html

A UK analytical-advisory service based on the BCSR approach is Glenside Fertility Farming Systems at:

http://www.glensideorganics.co.uk

A variety of soil tests are provided by Natural Resource Management Ltd, at:

http://www.nrm.uk.com/nrm_service.html

BCSR versus SLAN

The BCSR theory originated in New Jersey with Bear and co-workers (e.g. Bear and Prince, 1945; Bear and Toth, 1948). Following work with lucerne (*Medicago sativa* L.), they proposed an 'ideal ratio' of cation saturation of 65% Ca, 10% Mg, 5% K and 20% H. One of Albrecht's colleagues at the Missouri Agricultural Experiment Station was Dr E.R. Graham, who published an explanation of the BCSR theory and methods of soil testing (Graham, 1959). This theory was examined by researchers at the Department of Agronomy, Ohio State University, Columbus, and at the Ohio Agricultural Research and Development Centre, Wooster. Trials carried out, with the results appearing in three papers by McLean (1977), Eckert and McLean (1981) and McLean *et al.* (1983), and a chapter by Eckert (1987). They compared the BCSR approach with the sufficiency level of available nutrient (SLAN) viewpoint, which is credited to Bray (1944, 1945). The 1981 paper found that lucerne grew well at several ratios, and the data indicated that the balance of cations in the soil was unimportant, except at the extremely wide ratios where deficiencies of one element were caused by excesses of others. Hence no best ratio existed for Italian millet (*Setaria italica* (L.) Beauv.) or lucerne. The 1983 paper concluded that 'The results strongly suggest that for maximum crop yields, emphasis should be placed on providing sufficient, but non-excessive levels of each basic cation, rather than attempting to attain a favourable BCSR which evidently does not exist.' Weighing the two viewpoints, one can say that although BCSR values provide guidelines and broad limits of variation, recommendations must ultimately follow trials to determine the response of the crop in question under the expected field conditions, and one cannot be dogmatic about specific ratios. By 1987, Eckert concedes that both concepts can provide reasonable fertilizer recommendations if interpreted properly, however, the BCSR recommendations, unlike the SLAN, are not justified by agronomic research.

The Potash Development Association (PDA, 2000) reports that assessors in the US regarded the BCSR concept as most applicable to highly weathered soils of low pH requiring relatively major adjustments in fertility and where high Mg levels need to be maintained. As there are few such soils in the UK, it is questionable as to how applicable this concept is to the majority of UK soils.

For comments on BCSR versus SLAN for turfgrass, see the 1999 Iowa Turfgrass Research Report at the website:
http://www.hort.iastate.edu/pages/news/turfrpt/1999/calciumrod.html
Cation saturation ratios are also discussed in an article on the philosophy of soil testing in the National Corn Handbook, found at the website:
http://www.agcom.purdue.edu/AgCom/Pubs/NCH/NCH-2.html
The ineffectiveness of trying to achieve specific cation saturation ratios is presented in an article by Dr George Rehm (University of Minnesota) in the Wisconsin Crop Manager at:
http://ipcm.wisc.edu/wcm/99–14soils1.html

Phosphate Analysis

Organic farming methods may permit the application of natural rock phosphate, but not superphosphates. The former releases phosphate more slowly, therefore the extractant must be carefully chosen to reflect this fact. The acid Bray and Truog methods will extract too much phosphate to give spuriously high readings, whereas the alkaline extractants will extract very little. Preliminary tests indicate that the resin-extractable P method seems to give results which closely reflect the crop performance whether superphosphate or reactive phosphate rocks are used, and a discussion on the subject is given by Agricultural Consulting Services, Technical Services, at:

http://www.atsnet.co.nz/articles/resinp.htm

The Hislop and Cooke (1968) and Somasire and Edwards (1992) resin P methods are discussed in Chapter 4, and an automated method is given in 5.9c. 'Determination of resin extractable phosphorus (automated method)'.

Organic phosphorus

When superphosphate is applied to a subtropical soil, 40% may appear as organic P within 28 days of application (Dalal, 1977). It is also an accepted fact that from 30% to 85% of the total P of most soils is in organic combination, and that the phosphorus in manure is as available to plants as that in superphosphate (Sauchelli, 1965, pp. 79 and 195). The amount of P on a fresh-weight basis in cattle farmyard manure is approximately 1.54 kg t^{-1} (3.5 kg t^{-1} of P_2O_5) rising to 11.0 kg t^{-1} (25 kg t^{-1} P_2O_5) in broiler/turkey litter (Chambers *et al.*, 2001). When applied to the soil, soluble forms of phosphate are largely rendered insoluble by fixation to minerals, or immobilized by incorporation into microorganisms. Fixation may be by precipitation to form relatively insoluble forms of iron and aluminium phosphates or fluorapatite, or by chelation on to clay sesquioxides surfaces. Only a small amount remains in the soil solution, but the ability of the soil to maintain this plant-available P at an equilibrium level is the important factor. Although the microorganisms are initially more successful than plants at competing for soluble P, they serve to prevent the leaching of P, and when they die, mineralization of microbial P releases soluble P over a period of time to the benefit of the plants. There is therefore a delicate balance between immobilization and mineralization. The mineralization of acid soils is usually enhanced following liming, but not always. Some of the variation in the effect of lime may be produced by the Ca:Mg ratio effect on the mineralization and turnover of P in the soil. In alkaline soils, the ratios of organic C:organic P and total N:organic P should be greater than in acidic soils (Dalal, 1977).

Determination of organic P

The chemical nature of half of the soil organic P remains unknown, but there are three main groups: inositol phosphates, phospholipids and nucleic acids. There are also phosphoproteins and sugar phosphates (Dalal, 1977). There appears to be no direct methods for the determination of organic phosphorus in soils, but the indirect procedures of extraction and ignition may be used to fractionate the soil phosphorus approximately into inorganic, organic and total P. One of the main difficulties is choosing an extractant for inorganic P that will not simultaneously cause partial hydrolysis of the organic P fraction. One scheme of fractionation of extractable soil P has been presented by Williams (1950), and a method based on this is given in Method 11.1 below. This may be summarized as follows:

a = inorganic P (acetic acid extractable)
b = inorganic P (alkali extractable)
c = inorganic + organic P (alkali extractable)
$a + b$ = inorganic extractable P
$d = c - b$ = organic alkali extractable P
$a + b + d$ = total extractable P.

If required, the remaining P not extracted by either extractant is obtained by subtracting $(a + c)$ from the total P, which is determined by a separate P determination using a safer Na_2CO_3 fusion (Jackson, 1958) or the recommended, but more hazardous perchloric acid digestion technique (Olsen and Dean, 1965). An alternative ignition technique is given by Olsen and Dean (1965) whereby ignition converts organic P to inorganic P, and extraction of duplicate soil samples is carried out using concentrated HCl before and after ignition.

Method 11.1. Determination of extractable organic and inorganic soil P

Reagents.

- Acetic acid – 8-hydroxyquinoline reagent – dissolve 10 g 8-hydroxyquinoline in a solution of 2.5 % (v/v) acetic acid and make up to 1 l. (The 8-hydroxyquinoline blocks the readsorption or precipitation of phosphate by active iron and aluminium during acetic acid extraction. Synonyms: hydroxybenzopyridine; oxine; phenopyridine; 8-quinolinol. Not carcinogenic, but may be harmful if swallowed, and causes irritation to eyes, respiratory tract and skin; safety data sheet at http://www.jtbaker.com/msds/q7250.htm.)
- Hydrochloric acid solution, 1.0 M
- Magnesium nitrate, 0.5 M – dissolve 12.82 g $Mg(NO_3)_2.6H_2O$ in water and make up to 100 ml.
- Sodium hydroxide solution, 0.1 M

Procedure (extraction). Transfer 2.5 g air-dry soil, 2 mm mesh size, into a 250 ml polypropylene screw-cap centrifuge bottle/tube and add 100 ml acetic acid – 8-hydroxyquinoline reagent. Cap the tube and shake overnight (17 h) on a reciprocating shaker, at approximately 275 strokes of 25 mm length per minute at a constant temperature (20°C). Centrifuge for 15 min at 2800 rpm and remove an aliquot for the determination of acid extractable inorganic phosphorus (*a*).

Completely remove the remaining supernatant, dispense 100 ml 0.1 M sodium hydroxide into the tube and shake overnight as previously, being careful to maintain a constant temperature of 20°C, otherwise serious variations in the P extractability will occur. Centrifuge as above, and remove aliquots for determination of both the inorganic (*b*) and combined inorganic and organic fractions (*b* + *c*). If the presence of finely dispersed clay leaves a cloudy extract after centrifugation, dissolution of 0.5 g sodium chloride in the extract before centrifuging should result in a clear supernatant.

Procedure (determination of inorganic phosphate (a) in the acetic acid extract). The 8-hydroxyquinoline forms a precipitate in acidic ammonium molybdate solution, which will interfere unless the aliquot is <5 ml. It should therefore be removed by ignition as follows. Transfer 10 ml acetic acid extract to a 45-ml silica basin, add 0.5 ml 1 M magnesium acetate and evaporate to dryness on a water-bath. (*Note*: do not use magnesium nitrate, which reacts adversely on heating with 8-hydroxyquinoline.)

Transfer the basin to a cold muffle furnace and raise the temperature to a very dull red heat (545–555°C) and maintain for 20 min to oxidize all organic matter. Dissolve the residue in 15 ml M HCl and evaporate to dryness on a water-bath in a fume cupboard. Add 10 ml hot water and 1 ml 1 M HCl to dissolve the residue, then transfer with washings to a 100-ml beaker until about 40 ml. After cooling, adjust the pH to about 5.0 by dropwise addition of 1 M ammonium hydroxide, then transfer with rinsing to a 50-ml volumetric flask, make up to the mark and mix. Determine acetic acid extractable inorganic phosphate by a suitable procedure (see Methods 5.9a and 5.9b), noting that the results will be in mg P kg^{-1} soil. Note also that the solution obtained by the Williams procedure (2.5 g soil in 100 ml acetic acid solution, of which 10 ml is made up to 50 ml, i.e. a dilution of × 200 w/v) will be >10 × more dilute than using Olsen's bicarbonate extraction procedure (5 ml soil in 100 ml). Unless sufficient extra P is extracted by the acid compared with bicarbonate, a larger aliquot of the acetic acid extract may be necessary to provide adequate sensitivity for the determination unless the original (Truog and Meyer (1929); Truog (1930)) method is used. The calculation should take these dilutions into account.

Procedure (determination of inorganic phosphate (b) in the alkaline extract). Pipette 25 ml of the alkaline extract into a 50-ml volumetric flask followed by 3.5 ml 1 M HCl from a dispenser. This will neutralize the sodium hydroxide and leave 1 ml excess of acid, causing precipitation of organic matter. Make up to 50 ml and mix. Filter through a dry Whatman No. 44, 9 cm, filter paper into a dry 100-ml beaker, discarding the first few millilitres. Pipette a suitable aliquot (25–35 ml) into a 50-ml beaker, adjust the pH to approximately 5.0 as previously, make up to 100 ml and mix. Determine the inorganic phosphate (b) in the alkaline extract by a suitable procedure. Note that 2.5 g soil was extracted into 100 ml 0.1 M sodium hydroxide, of which a 25 ml aliquot was made up to 50 ml, of which 25 ml was made up to 100 ml. Thus 2.5 g soil has been extracted into the equivalent of 800 ml solution (× 320 w/v). This should be taken into account in the calculation.

Procedure (determination of inorganic plus organic phosphate (c) in the alkaline extract). Pipette a suitable aliquot (5 or 10 ml) of alkaline extract into a 45-ml silica basin, acidify with a few drops of concentrated nitric acid, add 2 ml 0.5 M magnesium nitrate and evaporate to apparent dryness on a water bath. Complete the evaporation gently on a hotplate in a fume cupboard, then transfer to a cold muffle furnace. Raise the temperature to a very dull red heat (545–555°C) and ignite for 20 min to oxidize all the organic matter and mineralize the organic P. Allow to cool, dissolve the residue in 15 ml of 1 M HCl, cover with a watch glass and heat for 30 min on a boiling water-bath. Remove the watch glass and rinse into the basin, then evaporate to dryness and redissolve the residue in 10 ml hot water plus 1 ml 1 M HCl. Filter through a 7-cm Whatman No. 41 filter paper into a 100-ml beaker. Wash the filter funnel and paper with hot water to bring the contents up to approximately 75 ml. Cool, adjust the pH to approximately 5.0, and transfer to a 100-ml volumetric flask with rinsing; make up to the mark and mix. Determine the inorganic plus organic phosphate (c) in the alkaline extract by a suitable procedure. Note that 2.5 g soil was extracted into 100 ml 0.1 M sodium hydroxide, of which a 5 or 10 ml aliquot was made up to 100 ml. There is therefore a × 400 or 800 w/v dilution factor to be incorporated in the calculation.

The Balzer Methods

A system of analytical methods of particular application to organic farming systems was created by Dr Friedrich M. Balzer in the early 1980s. The principles of his system are described in Balzer (1985), and is the methodology favoured by the Elm Farm Research Centre:

> http://www.efrc.co.uk/

also the Eco consultancy, Sustain-Ability, of Motueka, New Zealand:

> http://home.clear.net.nz/pages/awelte/ESTA.htm

An explanation of Balzer's method of interpretation has been given by Otto Schmid (Schmid, 1984).

Dr Balzer (e-mail: Dr.Balzer@labor-balzer.de) has kindly provided details of the sources of his analytical methods, and they are outlined below (F.M. Balzer, Wetter, Germany, 2001, personal communication). His laboratory website is:

> http://www.labor-balzer.de/index.htm

An English version of Balzer's advisory information and recommended levels of nutrients is found at:

> http://www.labor-balzer.de/lbz/e/index_e.htm

The soil is assessed using 14 tests, which cover physical, chemical, biological and environmental parameters, thus presenting a holistic view of the soil quality. Particular attention is given to humus dynamics and quality.

pH determination

The pH is determined in both water and 0.1 M KCl (10 g soil + 25 ml solution) as detailed in VDLUFA (1991, section A.5.1.1). *Note*: VDLUFA = Verband Deutscher Landwirtschaftlicher Untersuchungs- und Forschungsanstalten, whose website home page is:

> http://www.vdlufa.de/vdl_idx.htm

with method books listed at:

> http://www.vdlufa.de/vdl_2_2m.htm

From the home page go to LINK-Seite for addresses of useful agricultural organizations. Balzer says that from the relationship of the two pH values, additional information is given on the salt content and the exchange capacity of the soil. If the pH in water is only 0.1–0.3 higher than that in KCl, then the salt content of the soil is high; this may occur in intensively cultivated glasshouse soils and composts. If the pH difference exceeds 1.0 units, the salt and nutrient content will be low. The difference is usually about 0.5 for light soils, and 0.8–1.0 for medium to heavy soils with a high clay content. Following an additional calcium determination, the correct liming materials are recommended to maintain the optimum acid–base balance of the soil. Too much calcium will lock up the phosphorus, displace other cations such as magnesium, and reduce the availability of micronutrients such as boron, copper, iron, manganese and zinc.

Humus

Humus is equated with organic carbon, which is the carbon content, multiplied by 1.725, determined according to the Swiss Reference Methods of the Eidgenössenschaftliche landwirtschaftliche Forschunganstalten (Confederation of Agricultural Research Institutes) (FAW, 1998). For the Eidgenössische Forschungsanstalt für Obst-, Wein- und Gartenbau, Wädenswil (FAW) document search facility, see the website:

> http://www.admin.ch/sar/faw/docu/_suche_d.html

and for Swiss library holdings go to:

> http://candide.ethz.ch:4504/ALEPH/-/start/nebis-eng/new

Soils for crops and vegetables should contain at least 2% humus, and grassland over 5%.

Humification (see also Chapter 5, Method 5.4)

Apart from the overall humus content, the type of humus is important. Stable humus improves soil texture, and friable humus supplies nutritive substances to the plants. Balzer refers to Welte (1955), who used alkali and acid solutions to extract humic acids, which were classified into brown and grey humic acids. Paper electrophoresis gave two fractions for the brown and three for the grey humic acid. The optical densities at 472 (E4) and 664 nm (E6) are expressed as a ratio, termed a 'colour quotient' (Q 4/6), where Q 4/6 = E4/E6. Q 4/6 values for brown humic acids vary from 5.0 to 5.5, and grey humic acids from 2.2 to 2.8. The colour quotient of grey humic acid depends strongly on the nitrogen content, and will be displaced to lower values with increase in N-content. Further method details are in Schlichting and Blume (1961, pp. 126 and 136). There are others, however, who claim that 'There are no reliable or agreed scientific interpretations of the meanings of these ratio values' (Simpson *et al.*, 1997). Balzer gives the optimum humus content of sandy soils as 3.5–4.5%, loam as 3.0–4.0%, clay as 2.5–3.5% and peat bog as >20%.

Phosphorus, potassium and magnesium

These elements are determined at pH 3.6 according to the Egner-Riehm double-lactate extraction method (Egner *et al.*, 1960; VDLUFA, 1991; sections A.6.2.1.2 and A.6.2.4.2). The solubility of phosphorus in the soil is influenced by its biological activity and humus content. Soil microflora excrete organic acids such as acetic (weak), lactic (medium) and citric (strong), and these are chosen as extractants. The acetic acid-sodium acetate extraction is that of Morgan (Lund *et al.*, 1950), using 5 g soil plus 25 ml extractant at pH 4.8. Soil P is readily soluble in acetic acid, plant-available P is extracted by lactic acid, while the citric acid extraction indicates the potential reserves of P in the soil. This latter extractant is 2% citric acid, using 5 g soil plus 50 ml citric acid solution (VDLUFA, 1995). In a biologically active soil, the ratios of P extracted by acetic, lactic and citric acids should be about 1:3:9. Lower biological activity could widen the lactic:citric extractable P ratio to 2:10, and the biological processes should be stimulated by suitable organic fertilization using, e.g. green manures.

Potassium is determined in the acetic acid as well as the lactic acid extract. The potassium values for biologically managed fields lie between 100 and 200 mg K kg^{-1} soil, and are less than those recommended for conventional agriculture. Light soils normally have low values, and clay soils, which bind more potassium, have higher K levels. Potash fixation is also assessed according to Schlichting and Blume (1961, p. 84).

Trace elements

The elements Cu, Fe, Mn and Zn (also Ca) are analysed by the methods of Chapman and Pratt (1961) and also Perkins (1970). Perkins extracted 5 g air-dried soil (collected using an aluminium auger), sieved through a 2-mm stainless-steel mesh, with 20 ml of a solution of 0.05 M HCl plus 0.0125 M H_2SO_4 in a 250-ml large-mouth polythene bottle, and shaken on an Eberbach reciprocating shaker for 15 min at 280 oscillations per minute. It was filtered through a Whatman No. 44 paper through polypropylene funnels into polypropylene bottles (this should be modified to reject the first few millilitres of filtrate). The former extracting agent was originally chosen by Perkins because it was also used by many mid- and south-Atlantic states for extraction of P, K, Ca and Mg; whether it is appropriate in other locations should be determined. The detection of excessive amounts of Zn could arise from residues in sewage sludge or poultry manure. High Mn, or even Fe, values could arise from lack of aeration in medium to heavy soils.

The UK Situation

A recent review (Stockdale, 2001) concludes that:

1. There is insufficient knowledge appropriate to UK conditions and organic farming systems to allow organic farmers to make scientifically and economically sound management decisions enabling optimum sustainable use of P and K in organic systems.
2. Organic P is not routinely measured by any extraction procedure, and the conventional P index system cannot be simply applied in organic systems because of the complex interacting dynamics of the organic and mineral P pools in the soil.
3. Available K measured by ammonium nitrate gave a good indication of plant available K in the soil for organic systems, but not all extractable K is truly available. The K index system can be used as a reasonable guide for organic systems.

12 Quality Assurance and Control

Clearly, the commercial or consultancy laboratory that tests sub-samples of a marketed product worth millions of pounds, or assesses the purity of pharmaceuticals, or analyses forensic samples, must have far higher levels of both accuracy and verifiability than student practical classes. There should, however, always be an effort to produce the most accurate and reliable results within the constraints of the laboratory facilities available, otherwise a lax attitude will produce work of doubtful interpretation that could mislead others, as well as giving little job satisfaction. Several books, which are more suited to the commercial sector, have been written on the quality of laboratory analysis, however some quality assurance practices could be beneficial in the smaller laboratory. A useful open-learning style book on basic concepts of quality in the analytical laboratory has been co-authored by staff at the Laboratory of the Government Chemist (Crosby *et al.*, 1995).

The Laboratory of the Government Chemist (LGC) website is at:

      http://www.lgc.co.uk

Two important definitions are those of *quality control* and *quality assurance*. The former relates to operational techniques and activities, whereas the latter ensures that systematic actions are in place which enable confidence that the results meet the required level of quality, such as accuracy and precision. The concept of *total quality control* extends to areas such as management style and reduction of waste.

Quality control would include the following:

1. Analysis of replicates to determine precision;

200   © 2002 CAB *International. Methods in Agricultural Chemical Analysis: a Practical Handbook* (N.T. Faithfull)

2. Blank samples to detect impurities in the reagents or interferences;
3. Standard reference materials to check the accuracy of the method.

Replicates

If a certain analysis is frequently carried out, it may be advantageous to keep in stock some well-mixed bulk samples (low, medium and high range values), sub-samples of which should be included with every fresh batch of analyses, and again well-mixed before weighing. After a statistically significant number of analyses, an accepted average value for the bulk sample is obtained, and the amount of scatter of results computed as a standard deviation, s, where

$$s = \sqrt{\frac{\sum\left(x - \bar{x}\right)^2}{n-1}}$$

n = total number of results
x = observed value
$\bar{x}$ = mean value of observed concentrations
$(x - \bar{x})$ = deviation from mean

If subsequent analyses of the bulk sample deviate by more than a predetermined amount, the whole batch of results is rejected. Results are thus only accepted if they fall between specified values of s above and below the mean, where $1s$ includes 68%, $2s$ includes 95% (the normally accepted value), and $3s$ includes 99.7% of results. The scatter of results usually assumes a symmetrical normal or Gaussian distribution about the mean, as shown in Figs 12.1 and 12.2.

Bulk samples are repeated, and if still outside the acceptable limits of precision, the methodology must be examined for sources of error; this was considered fully in an early paper by Büttner (1968).

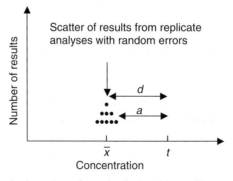

Fig. 12.1. Typical scatter of results from nine replicate analyses. a = absolute error of the determination, d = systematic error, n = total number of results (=9 in Fig. 12.1; see under 'Replicates' in the text), t = true value, $\bar{x}$ = mean observed concentration value, x = observed value (see under 'Replicates' in text).

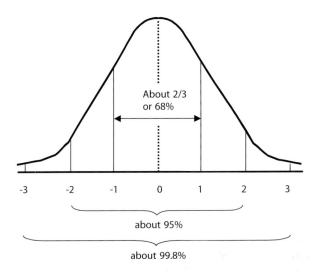

About 2/3
or 68%

-3   -2   -1   0   1   2   3

about 95%

about 99.8%

Fig. 12.2. A normal or Gaussian distribution of results with % population enclosed by various ± standard deviation values.

With automated segmented flow analysis, the scatter (or distribution) of results often departs from a normal distribution, and may be skewed (Faithfull, 1972). The tendency is for results over a 90-min period (sampling rate 40 h⁻¹) to be negatively skewed, with a tail at lower values and the peak occurring at a higher value. This probably results from changes in the flexing properties of the pump tubing, with 20–40 min of reagent pumping required before an approximately normal distribution of results with an acceptable standard deviation is obtained. Acidflex acid-resistant tubing has been shown to require up to 1 h to stabilize (Davidson *et al.*, 1970). The question of calibration drift and specimen interaction in segmented flow analysis was discussed by Bennet *et al.* (1970).

Odd sample values occurring some way from the cluster of values around the mean are known as outliers. The problem of whether or not they are acceptable, especially with skewed distributions of data, is considered in AMC (2001), which is also found at:

http://www.rsc.org/lap/rsccom/amc/amc_index.htm

Outlier tests, and other statistical methods in analytical chemistry, are also discussed by Meier and Zünd (1993).

Standard Reference Materials

Standard reference materials are available from several suppliers; in the UK, a selection of agricultural substances is available from the Laboratory of the Government Chemist, and in the US from the National Institute of Standards and Technology (NIST), where the Standard Reference Materials programme is described at:

http://ts.nist.gov/ts/htdocs/230/232/232.htm

Quality Systems

Quality systems are systems that ensure that both the laboratory and its procedures, and the staff and overall management, together provide an overall quality of service. This necessitates regular reviews to check the maintenance of this quality. There are organizations that define standards to be met and issue certificates to qualifying laboratories. These are subject to repeated inspections, and the cost must be reflected in the charges for analyses. Such bodies are the Organisation for Economic Co-operation and Development (OECD) which has developed the Good Laboratory Practice (GLP) standard; the International Organization for Standardization has produced the ISO 9000 series of standards applicable to laboratories, with a web page at:

> http://www.iso.ch/iso/en/ISOOnline.frontpage

The LGC promotes best practice in valid analytical measurement, and the principles are detailed at:

> http://www.vam.org.uk/aboutvam/about_principles.asp

with details on proficiency testing and links to other organizations accessible from the home page:

> http://www.vam.org.uk/

Two UK proficiency testing schemes are the Food Analysis Performance Assessment Scheme (FAPAS), which replaced NAMAS in 2000, which is arranged by the Central Science Laboratory, with details at:

> http://ptg.csl.gov.uk/fapas.cfm

and the Food Examination Performance Assessment Scheme (FEPAS) for microbiological proficiency assessment. Proficiency testing (interlaboratory comparison programmes) with respect to chemical analysis involves the submission of a sample of the relevant material with validated known attributes, prepared in an accredited laboratory, to a comparative analysis in the subject laboratory. The results are then evaluated, either statistically or analytically and returned as a score to the subject laboratory. Any poor results are investigated to determine the cause of the inaccuracy, and the methodology is adjusted accordingly until the required degree of repeatability and reproducibility has been achieved (Charlett, 1996). Sometimes the customer's fitness-for-purpose criterion is different from that set by the proficiency testing scheme. One solution is for participants to calculate an auxiliary score called the 'zeta-score'. This is beyond the scope of this chapter, but details may be consulted in AMC (2000) and at:

> http://www.rsc.org/lap/rsccom/amc/amc_index.htm

Certain organizations will help laboratories to achieve accreditation, an example being the United Kingdom Accreditation Service (UKAS):

> http://www.ukas.com

In the US, the NIST administers the National Voluntary Laboratory Accreditation Program (NVLAP), which complies with ISO 9002, and with details available at:

> http://ts.nist.gov/ts/htdocs/210/214/summary.htm

Matrix Interference

One often unsuspected source of error can arise from interference by the substances originating in the sample which are present in addition to the analyte, and which are collectively termed the matrix. The matrix components could enhance, diminish or have no effect on the measured reading, when present within the normal range of concentrations. Atomic absorption spectrophotometry is particularly susceptible to this type of interference, especially with electrothermal atomization. Flame AAS may also be affected by the flame emission or absorption spectrum, even using ac modulated hollow cathode lamp emission and detection (Faithfull, 1971b, 1975).

Standard Additions, or 'Spiking'

Ideally, the standards should be made up in a solution containing the same normally expected levels of matrix elements as occur in the sample solution. It should be borne in mind that even if they exert no chemical interference, they could possibly exert a viscosity effect on a nebulized solution (especially with high concentrations of phosphoric or sulphuric acids). If it is not possible to determine the matrix components or prepare standards in a matrix solution, and unless experiments have shown matrix interference to be insignificant, then the method of standard additions, or 'spiking', should be carried out. This is where known amounts of the analyte are added to the sample or sample solution before determination by, e.g. AAS or colorimetry.

There are several methods of carrying out the standard addition method. They all rely, however, on the standard curve being linear over the range of analyte concentrations in the spiked solutions. The addition of an incremental range of standards, rather than just one, will improve precision. One method (Thomas, 1996) involves pipetting a fixed volume (V_x) of sample solution into a series of volumetric flasks, next adding a range (say, 0, 5, 10, 15 and 20 ml) of a standard analyte solution (concentration C_s) to the respective flasks, followed by the reagents, then making up to the mark. The absorbance (y-axis) is plotted against volume of standard added. If α is the intercept on the y-axis, and β is the gradient, then the concentration of the analyte in the unknown solution (C_x) is given by:

$$C_x = \alpha C_s / \beta V_x$$

Two other approaches are given by Meier and Zünd (1993). The observed signal is plotted against the amount of analyte spiked (e.g. mg ml^{-1} in the directly analysed solution) into the test sample (two or more points). The regression line is extrapolated to $y = 0$, and the unknown concentration is given by $-1 \times$ intercept on the x-axis (which is a negative value). In certain circumstances, e.g. where interference by a matrix component is present, the line is extrapolated to a level where $y > 0$. The other approach is to use an interpolation rather than an extrapolation, which improves precision with no additional tests. The method is to subtract the reading of the unspiked sample from each of the spiked sample readings. The difference is plotted as the best straight line passing through the origin ($x = 0$; $y = 0$). The concentration value corresponding to the reading for the unspiked sample is read from the standard addition line.

References

ADAS (1987) *The Use of Sewage Sludge on Agricultural Land*. Booklet 2409. MAFF Publications, Alnwick.

Adriano, D.C., Paulsen, G.M. and Murphy, L.S. (1971) Phosphorus–iron and phosphorus–zinc relationships in corn (*Zea mays* L.) seedlings as affected by mineral nutrition. *Agronomy Journal* 63(1), 36.

Aiken, G.R., McKnight, D.M., Wershaw, R.L. and MacCarthy, P. (1985) An introduction to humic substances in soil, sediment and waters. In: Aiken, G.R., MacCarthy, P., Malcolm, R.L. and Swift, P.S. (eds) *Humic Substances in Soil, Sediment and Water*. John Wiley & Sons, New York, pp. 1–9.

Albrecht, W.A. (1942) Buying fertilizers wisely. *Missouri Agricultural Experiment Station Circular 227*, April 1942.

Albrecht, W.A. (1960) Boron interrelated with potassium. *Let's Live*, May 1960.

Albrecht, W.A. (1963) Organic matter balances soil fertility. *Natural Food and Farming*, February 1963.

Albrecht, W.A. (1967) Soil reaction (pH) and balanced plant nutrition. Mimeographed booklet distributed by the author, 38 pp.

Alderman, G. and Cottrill, B.R. (Compilers) (1993) *Energy and Protein Requirements of Ruminants*. CAB International, Wallingford, UK, 159 pp.

Alexander, R.H. (1960) The sampling of silage pits by coring. *Journal of Agricultural Engineering Research* 5, 118–122.

Allen, M. and Whitfield, A.B. (1964) Rapid methods for the routine determination of major nutrient elements and iron and manganese in the leaves of fruit trees, *Annual Report of the East Malling Research Station*, pp. 143–147.

Allison, L.E. (1960) Wet combustion apparatus and procedure for organic and inorganic carbon in soil. *Soil Science Society of America Proceedings* 24, 36–40.

AMC (2000) How to combine proficiency test results with your own uncertainty estimate – the zeta score. *AMC Technical Brief No. 2* (Nov 2000). Analytical Methods Committee, Royal Society of Chemistry, London, 2 pp.

AMC (2001) Robust statistics: a method of coping with outliers. *AMC Technical Brief No. 6* (April 2001). Analytical

Methods Committee, Royal Society of Chemistry, London, 2 pp.

[ARC] Agricultural Research Council (1963) *Report of Group on Comparison of Methods of Analysis of Mineral Elements in Plants*. Agricultural Research Council, London, 52 pp.

Archer, F.C. (1980) Trace elements in soils in England and Wales. In: *Inorganic Pollution and* Agriculture, MAFF/ADAS Reference Book 326, pp. 184–190.

Archer, F.C. and Hodgson, I.H. (1987) Total and extractable trace element contents of soils in England and Wales. *Journal of Soil Science* 38, 421–431.

Ayres, J.F. (1991) Sources of error with *in vitro* digestibility assay of pasture feeds. *Grass and Forage Science* 46, 89–97.

Armitage, E.R. (1965) Use of the AutoAnalyzer in an advisory soil service. Technicon Fifth International Symposium, 'Automation in Analytical Chemistry,' London, 13 Oct. 1965, Pub. 1966, Mediad Inc., pp. 142–143.

Baker, A.S. and Smith, R. (1969) Extracting solution for potentiometric determination of nitrate in plant tissue. *Journal of Agricultural and Food Chemistry* 17(6), 1284–1287.

Baker, C.W. and Givens, D.I. (1992) The use of near infra-red reflectance spectroscopy (NIRS) for the evaluation of forages: the application to undried silage. *Animal Production* 54, 507 (Abstract).

Balzer, F.M. (1985) Dr. Balzer's soil analysis method. *Elm Farm Research Centre: Soil Analysis Service – Research Notes No. 4*, pp. 4–11 [Bodenanalyse-System Dr. Balzer, *Lebendige Erde*, Jan.-Feb. issue (1984), Verlagsort Darmstadt, pp. 13–18, translated by Nic Lampkin].

Barber, G.D., Givens, D.I., Kridis, M.S., Offer, N.W. and Murray, I. (1990) Predicting the organic matter digestibility of grass silage. *Animal Feed Science and Technology* 28, 115–128.

Barker, A.V., Peck, N.H. and MacDonald, G.E. (1971) Nitrate accumulation in vegetables. I. Spinach grown in upland soils. *Agronomy Journal* 63, 126–129.

Barnes, R.J., Dhanoa, M.S. and Lister, S.J.

(1989) Standard normal variate transformation and detrending of near infrared diffuse reflectance spectra. *Applied Spectroscopy* 43, 772–777.

Barton, F.E., II, Shenk, J.S., Westerhaus, M.O. and Funk, D.B. (2000) The development of near infrared wheat quality models by locally weighted regressions. *Journal of Near Infrared Spectroscopy* 8, 201–208.

Batten, G.D. and Blakeney, A.B. (1992) The determination of sulphur in plants using NIR. In: Hildrum, K.I., Isaksson, T., Naes, T. and Tandberg, A. (eds) *Near Infra-red Spectroscopy*. Ellis Horwood, New York, pp. 185–190.

Bear, F.E. and Prince, A.L. (1945) Cation equivalent constancy in alfalfa. *Journal of the American Society of Agronomy* 37, 217–222.

Bear, F.E. and Toth, S.J. (1948) Influence of calcium on availability of other soil cations. *Soil Science* 65, 67–74.

Bennet, A., Gartelmann, D., Mason, J.I. and Owen, J.A. (1970) Calibration, calibration drift and specimen interaction in autoanalyser systems. *Clinica Chimica Acta* 29, 161–180.

Berardo, N. (1992) Measuring Italian ryegrass quality by near infrared reflectance spectroscopy (NIRS). In: Murray, I. and Cowe, I.A. (eds) *Making Light Work: Advances in Near Infrared Spectroscopy*. VCH, Weinheim, Germany, pp. 272–276.

Berrow, M.L. (1988) Sampling of soils and plants for trace element analysis. *Analytical Proceedings* 25 (4) 116–118.

Berrow, M.L. and Burridge, J.C. (1980) Trace element levels in soil: effect of sewage sludge. In: *Inorganic Pollution and Agriculture*, MAFF/ADAS Reference Book 326. HMSO, London, pp. 159–183.

Berthelot, M.P. (1859) Violet d'analine. *Repertoire de Chimie Applique* p. 284.

Bertoldi, M. de, Ferranti, M.P., L'Hermite, P. and Zucconi, F. (1987) *Compost: Production, Quality and Use*. Elsevier, London, 853 pp.

Bertone, A.L., Van Soest, P.J., Johnson, D., Ralston, S.L. and Stashak, T.S. (1989a) Large intestinal capacity, retention times, and turnover rates of particulate ingesta associated with extensive large colon resection in horses. *American Journal of Veterinary Research* 50, 1621–1627.

Bertone, A.L., Ralston, S.L. and Stashak, T.S. (1989b) Fibre digestion and voluntary intake in horses after adaptation to extensive large colon resection. *American Journal of Veterinary Research* 50, 1628–1632.

Bettany, J.R. and Helstead, E.H. (1972) An automated procedure for the nephelometric determination of sulfate in soil extracts. *Canadian Journal of Soil Science* 52, 127–129.

Bingham, F.T. (1982) Boron. In: Page, A.L. (ed.) *Methods of Soil Analysis*, Part 2, 2nd edn. Soil Science Society of America, Madison, Wisconsin, pp. 431–447.

Blaxter, K.L. and Clapperton, J.L. (1965) Prediction of the amount of methane produced by ruminants. *British Journal of Nutrition* 19, 511–522.

Blaxter, K.L., Graham, N.Mc. and Wainman, F.W. (1956) Some observations on the digestibility of food by sheep and on related problems. *British Journal of Nutrition* 10, 69–91.

Boer, F. De. and Bickel, H. (eds) (1988) *Livestock Feed Resources and Feed Evaluation in Europe.* (European Association for Animal Production, Publication No. 37), Elsevier, Amsterdam, pp. 115–121.

Bonanno, A.S., Olinger, J.M. and Griffiths, P.R. (1992) In: Hildrum, K.I., Isaksson, T., Naes, T. and Tandberg, A. (eds) *Near Infra-red Spectroscopy.* Ellis Horwood, New York.

Bradfield, R. (1942) Liebig and the chemistry of the soil. In: Moulton, F.R. (ed.) *Liebig and After Liebig: a Century of Progress in Agricultural Chemistry.* American Association for the Advancement of Science, Washington, DC, pp. 48–55.

Bray, R.H. (1944) Soil–plant relationships: I. The quantitative relation of exchangeable K to crop yields and to crop response to potash additions. *Soil Science* 58, 305–324.

Bray, R.H. (1945) Soil–plant relationships: II.

Balanced fertilizer use through soil tests for K and P. *Soil Science* 60, 463–473.

Bray, R.H. and Kurtz. (1945) Determination of total, organic and available form of phosphorus in soils. *Soil Science* 59, 39–45.

Bremner, J.M. and Keeney, D.R. (1965) Steam distillation methods for determination of ammonium, nitrate and nitrite. *Analytica Chimica Acta* 32, 485–495.

Broderick, G.A. and Cochran, R.C. (2000) *In vitro* and *in situ* methods for estimating digestibility with reference to protein degradability. In: Theodorou, M.K. and France, J. (eds) *Feeding Systems and Feed Evaluation Models.* CAB International, Wallingford, UK, pp. 53–85.

Browne, C.A. (1942) Liebig and the Law of the Minimum. In: Moulton, F.R. (ed.) *Liebig and After Liebig: A Century of Progress in Agricultural Chemistry.* American Association for the Advancement of Science, Washington, DC, pp. 71–82.

BS 5766 (1979) *Analysis of Animal Feeding Stuffs, Part 1, Determination of Crude Ash.* British Standards Institute, London. ISO 5984–1978, International Organization for Standardization, Geneva.

Burns, D.T. (1984) Kjeldahl, the man, the method and the Carlsberg laboratory. *Analytical Proceedings* 21(6), 210–214.

Büttner, H. (1968) Qualitätskontrolle im automatisierten klinisch-chemischen Laboratorium. *Zeitschrift für Chemie und klinische Biochemie* 7 (1), 89–95.

Byrne, E. (1979) *Chemical Analysis of Agricultural Materials.* An Foras Taluntais, Wexford, Eire, p. 52.

Canale, A., Valente, M.E. and Ciotti, A. (1984) Determination of volatile carboxylic acids (C_1–C_{5i}) and lactic acid in aqueous extracts of silage by high performance liquid chromatography. *Journal of the Science of Food and Agriculture* 35, 1178–1182.

Chambers, B., Nicholson, N., Smith, K., Pain, B., Cumby, T. and Scotford, I. (2001) *Managing Livestock Manures: Making Better Use of Livestock Manures on Arable Land*, Booklet 1, 2nd edn. ADAS Gleadthorpe Research Centre, Mansfield, UK, 25 pp.

Chapman, H.D. and Pratt, P.F. (1961) *Methods of Analysis for Soils, Plants and Waters,* Priced Publication 4034. University of California, Division of Agricultural Sciences, Berkeley, 309 pp.

Charlett, S.M. (1996) Proficiency testing and the smaller food laboratory. *VAM Bulletin* 14, 12–13.

Cherian, G., Sauer, W.C. and Thacker, P.A. (1988) Effect of predigestion factors on the apparent digestibility of protein for swine determined by the mobile nylon bag technique. *Journal of Animal Science* 66 (8), 1963–1968.

Cherian, G., Sauer, W.C. and Thacker, P.A. (1989) Factors affecting the apparent digestibility of protein for swine when determined by the mobile nylon bag technique. *Animal Feed Science and Technology* 27(1–2), 137–146.

Cherney, D.J.R. (2000) Characterization of forages by chemical analysis. In: Givens, D.I., Owen, E., Axford, R.F.E. and Omed, H.M. (eds) *Forage Evaluation in Ruminant Nutrition.* CAB International, Wallingford, UK, pp. 281–300.

Chesson, A. (2000) Feed characterization. In: Theodorou, M.K. and France, J. (eds) *Feeding Systems and Feed Evaluation Models.* CAB International, Wallingford, UK, pp. 11–33.

Christian, G.D. and Feldman, F.J. (1970) *Atomic Absorption Spectroscopy: Applications in Agriculture, Biology and Medicine.* Wiley Interscience, New York, pp. 118–119.

Clancy, M.J. and Wilson, R.K. (1966) Development and application of a new chemical method for predicting the digestibility and intake of herbage samples. *Proceedings of the Xth International Grassland Congress,* Helsinki, 445–452.

Coleman, S.W., Lippke, H. and Gill, M. (1999) Estimating the nutritive potential of forages. In: Jung, H.J.G. and Fahey, G.C. Jr, (eds) *Nutritional Ecology of Herbivores,* Proceedings of Vth International Symposium on Nutrition of Herbivores. American Society of Animal Science, pp. 647–695.

Combs, D.K and Satter, L.D. (1992) Determination of markers in digesta and feces by direct current plasma emission spectroscopy. *Journal of Dairy Science* 75, 2176–2183.

Combs, S.M., Denning, J.L. and Frank, K.D. (1998) Sulfate-sulfur. In: *Recommended Chemical Soil Test Procedures for the North Central Region.* NCR Research Publication No. 221 (Revised), Missouri Agricultural Experiment Station, pp. 35–40.

Committee on Chemical Safety (1995) *Safety in Academic Chemistry Laboratories,* 6th edn. American Chemical Society, Washington, DC, 70 pp.

Cone, J.W. (1997) The development, use and application of the gas production technique at ID-DLO. *Proceedings of the British Society of Animal Science International Symposium on In Vitro Techniques for Measuring Nutrient Supply to Ruminants.* University of Reading, Whiteknights, Reading, UK, p. 9.

Cork, S.J., Hume, I.D. and Faichney, G.J. (1999) Digestive strategies of nonruminant herbivores: the role of the hindgut. In: Jung, H.-J.G. and Fahey, G.C., Jr (eds) *Nutritional Ecology of Herbivores: Proceedings of the Vth International Symposium on the Nutrition of Herbivores.* American Society of Animal Science, Savoy, Illinois, pp. 210–260.

Coutinho, J. (1997) Automated method for sulphate determination in soil-plant extracts and waters. In: Hood, T.M. and Benton Jones, J., Jr (eds) *Soil and Plant Analysis in Sustainable Agriculture and Environment.* Marcel Dekker, New York, pp. 481–494.

Craven, P. and Schwer, E.W. (1960) The determination of phosphorus. In: *Joint Symposium on Fertiliser Analysis.* Proceedings No. 62, The Fertiliser Society, London, pp. 139–175. Discussion, pp. 176–183.

Cresser, M. (1990) Sample preparation in environmental chemistry. *Analytical Proceedings* 27(5), 110–111.

Crosby, N.T., Day, J.A., Hardcastle, W.A., Holcombe, D.G. and Treble, R.D. (1995) *Quality in the Analytical Chemistry*

Laboratory. John Wiley & Sons (ACOL Series), Chichester, UK, 307 pp.

Curtis, H.A. (1942) Liebig and the chemistry of mineral fertilizers. In: Moulton, F.R. (ed.) *Liebig and After Liebig: A Century of Progress in Agricultural Chemistry.* American Association for the Advancement of Science, Washington, DC, pp. 64–70.

Dalal, R.C. (1977) Soil organic phosphorus. *Advances in Agronomy* 29, 83–117.

Dart, R.K. (1996) *Microbiology for the Analytical Chemist.* The Royal Society of Chemistry, Cambridge, UK, 159 pp.

Davidson, J., Mathieson, J. and Boyne, A.W. (1970) The use of automation in determining nitrogen by the Kjeldahl method, with final calculations by computer. *Analyst* 95, 181–193.

Deaville, E.R. and Baker, C.W. (1993) Spectral interpretations in relation to food characterisation. Paper 3 In: *NIR Spectroscopy: Developments in Agriculture and Food,* Proceedings of the International Conference on NIR Spectroscopy. ADAS, Drayton, UK, pp. 1–20.

Deaville, E.R. and Givens, D.I. (1996) Prediction of the non-fermentable energy fraction of grass silage using near infrared reflectance spectroscopy. Poster No. 218, the 112th meeting of the British Society for Animal Science, Scarborough, 18–20 March, 1996. *Animal Science* 62, 681.

de la Roza, B. and Martinez, A. (1992) The use of near infrared reflectance spectroscopy to predict the nutritive value and *in vivo* digestibility of grass silages. In: Murray, I. and Cowe, I.A. (eds) *Making Light Work: Advances in Near Infrared Spectroscopy.* VCH, Weinheim, Germany, pp. 269–271.

Delaware Cooperative Extension (1995) *Recommended Soil Testing Procedures for the Northeastern United States,* 2nd edn. Northeastern Regional Publication No. 493, Delaware Cooperative Extension System, University of Delaware, Newark.

Derrick, R.W., Moseley, G. and Wilman, D. (1993) Intake, by sheep, and digestibility of chickweed, dandelion, dock, ribwort and spurrey, compared with perennial ryegrass. *Journal of Agricultural Science,* 120, 51–61.

Dewar, W.A. and McDonald, P. (1961) Determination of dry matter in silage by distillation with toluene. *Journal of the Science of Food and Agriculture* 12, 790–795.

Dhanoa, M.S. (1988) On the analysis of dacron bag data for low degradability feeds. *Grass and Forage Science* 43, 441–444.

DOE/NWC (1981) *Report of the sub-committee on the disposal of sewage sludge to land.* Standing Technical Committee Report No. 20. Department of the Environment, London.

Duke, J.A. (2000a) *Handbook of Phytochemical Constituents of GRAS Herbs and Other Economic Plants: Herbal Reference Library.* CRC Press, Boca Raton, Florida, 680 pp.

Duke, J.A. (2000b) *Handbook of Medicinal Herbs: Herbal Reference Library.* CRC Press, Boca Raton, Florida, 696 pp.

EC (1992) Determination of crude fibre. *Official Journal of the European Communities* L344, 35–37 (26.11.92).

Eckert, D.J. (1987) Soil test interpretations: Basic cation saturation ratios and sufficiency levels. In: Brown, J.R. (ed.) *Soil Testing: Sampling, Correlation, Calibration, and Interpretation.* Soil Science Society of America Special Publication Number 21, Madison, Wisconsin, pp. 53–64.

Eckert, D.J. and McLean, E.O. (1981) Basic cation saturation ratios as a basis for fertilizing and liming agronomic crops: I. Growth chamber studies. *Agronomy Journal* 73, 795–799.

Englyst, H.N. and Cummings, J.H. (1984) Simplified method for the measurement of total non-starch polysaccharides by gas–liquid chromatography of constituent sugars as alditol acetates. *Analyst* 109, 937–942.

Egner, H., Riehm, H. and Domingo, W. (1960) Untersuchungen über die chemische Bodenanalyse als Grundlage für

Beurteilung des Nährstoffzustandes der Böden. *Annals of the Royal Agricultural College of Sweden* 26, pp. 199 et ff.

Estrada, J., Sana, J., Cequeil, R.M. and Cruanas, R. (1987) Application of a new method for CEC determination as a compost maturity index. In: De Bertoldi, M., Ferranti, M.P., L'Hermite, P., and Zucconi, F. (eds) *Compost: Production, Quality and Use.* Elsevier Applied Science, London, pp. 334–340.

Faithfull, N.T. (1969) Multiple Kjeldahl digestion unit. *Laboratory Practice* 18(12), 1302.

Faithfull, N.T. (1970) Automated methods in herbage analysis. MSc. thesis, University of Wales, pp. 32–36.

Faithfull, N.T. (1971a) Automated simultaneous determination of nitrogen, phosphorus, potassium and calcium on the same herbage digest solution. *Laboratory Practice* 20(1), 41–44.

Faithfull, N.T. (1971b) Flame interference in atomic absorption spectroscopy with a.c. modulated systems. *Laboratory Practice* 20(8), 641–643.

Faithfull, N.T. (1972) Variation of precision with time for AutoAnalyzer data. *Laboratory Practice* 21(4), 245–247 and 268.

Faithfull, N.T. (1974) Conversion of the Technicon Model II flame photometer to premix burner. *Laboratory Practice,* 23(8), 429–430.

Faithfull, N.T. (1975) Spectrum of the acetylene/air flame used in the determination of iron by atomic absorption spectrophotometery. *Laboratory Practice* 24(11), 737–739.

Faithfull, N.T. (1984) The in-vitro digestibility of feedstuffs – a century of ferment. *Journal of the Science of Food and Agriculture* 35, 819–826.

Faithfull, N.T. (1990) Acid hydrolysis prior to automatic analysis for starch. *Journal of the Science of Food and Agriculture* 50, 419–421.

Faithfull, N.T. (1993) Agricultural chemistry and research institutions: past, present and prospective. *Journal of the University of Wales Agricultural Society* 73, 116–125.

Faithfull, N.T. (1996) NIR – a blackbox technique for agricultural analyses. *Journal of the University of Wales Agricultural Society* 76, 29–38.

Faithfull, N.T. (1997) Problems and progress in silage sampling and analysis. *Journal of the University of Wales Agricultural Society* 77, 55–76.

Faithfull, N.T. (1998) Determination of dry matter in small silage samples. *Journal of the University of Wales Agricultural Society* 78, 101–114.

FAW (1998) Schweizerische Referenzmethoden der Eidgenössenschaftliche landwirtschaftlichen Forschunganstalten. (Authored by various institutions: Eidgenössische Forschungsanstalt für Obst-, Wein- und Gartenbau, Wädenswil (FAW), etc.), Volume 1, Published 1996, amended 1998.

Field, M. (1995) Interpretation of near infrared spectra of silages in relation to predicted and actual feeding value. Ph.D. thesis, University of Wales, Aberystwyth, 336 pp.

Finck, A. (1982) *Fertilizers and Fertilization.* Verlag Chemie, Weinheim, Germany, 438 pp.

Flaig, W. (1997) Aspects of the biochemistry of the healing effects of humic substances from peat. In: Hayes, M.H.B. and Wilson, W.S. (eds) *Humic Substances in Soils, Peats and Waters.* The Royal Society of Chemistry, Cambridge, UK, pp. 346–356.

Fogg, D.N. and Wilkinson, N.T. (1958) The colorimetric determination of phosphorus. *Analyst* 83, 406–414.

Follett, M.J. and Ratcliff, P.W. (1963) Determination of nitrite and nitrate in meat products. *Journal of the Science of Food and Agriculture* 14, 138–144.

Frape, D. (1986) *Equine Nutrition and Feeding.* Longman Scientific and Technical, Harlow, UK.

Fuller, C.W. (1977) *Electrothermal Atomization for Atomic Absorption Spectrometery.* The Chemical Society, London, pp. 56–58.

Furr, A.K. (ed.) (2000) *CRC Handbook of Laboratory Safety,* 5th edn. CRC Press, Boca Raton, Florida, 808 pp.

Fussell, R.J. and McCalley, D.V. (1987)

Determination of volatile fatty acids (C_2–C_5) and lactic acid in silage by gas chromatography. *Analyst* 112, 1213–1216.

Galletti, G.C. and Piccaglia, R. (1983) Rapida analisi gascromatografica dell'acido lattico e degli acidi organici volatili nel silomais. *Zootecnica e Nutrizione Animale* 9, 347–356.

Gehrke, C.W., Ussary, J.P. and Kramer, G.H. Jr (1964) Automation of the AOAC flame photometric method for potassium in fertilizers. *Journal of the Association of Official Agricultural Chemists* 47, 459–469.

Givens, D.I. (1993) The role of NIRS for forage evaluation – the present and future. In: *NIR Spectroscopy: Developments in Agriculture and Food*, Proceedings of the International Conference on NIR Spectroscopy, ADAS, Drayton, UK, 19 pp.

Givens, D.I., Everington, J.M. and Adamson, J.H. (1989) The digestibility and ME content of grass silage and their prediction from laboratory measurements. *Animal Feed Science Technology* 24, 27–43.

Givens, D.I., Everington, J.M. and Adamson, J.H. (1990) The nutritive value of Spring-grown herbage produced on farms throughout England and Wales over 4 years. III. The prediction of energy values from various laboratory measurements. *Animal Feed Science Technology* 27, 185–196.

Givens, D.I., Owen, E. and Adesogan, A.T. (2000) Current procedures, future requirements and the need for standardisation. In: *Forage Evaluation in Ruminant Nutrition*, CAB International, Wallingford, UK, pp. 449–474.

Gomez, K.A. and Gomez, A.A. (1984) *Statistical Procedures for Agricultural Research*, 2nd edn. John Wiley & Sons, New York, 680 pp.

Grace, R. and Mirna, A. (1957) *Fresenius' Zeitschrift für Analytische Chemie* 158, 182–189.

Graham, E.R. (1959) An explanation of theory and methods of soil testing. *Missouri Agricultural Experiment Station Bulletin* No. 734, University of Missouri-Columbia.

Graham, H. (1991) The physical and chemical constitution of foods: effects on carbohydrate digestion. In: Fuller, M.F. (ed.) In vitro *Digestion for Pigs and Poultry*. CAB International, Wallingford, UK, pp. 35–44.

Hall, M.B., Jennings, J.P., Lewis, B.A. and Robertson, J.B. (2001) Evaluation of starch analysis methods for feed samples. *Journal of the Science of Food and Agriculture* 81, 17–21.

Hall, M.B., Lewis, B.A., Van Soest, P.J. and Chase, L.E. (1997) A simple method for estimation of neutral detergent-soluble fibre. *Journal of the Science of Food and Agriculture* 74, 441–449.

Harada, Y. and Inoko, A. (1980) The measurement of the cation exchange capacity of composts for the estimation of the degree of maturity. *Soil Science and Plant Nutrition* 26, 127–134.

Harada, Y., Inoko, A., Tadaki, M. and Izawa, T. (1981) Maturing process of city refuse compost during piling. *Soil Science and Plant Nutrition* 27, 357–364.

Harborne, J.B. (1984) *Phytochemical Methods*. Chapman and Hall, London, pp. 44–45. (Reprinted 1991.)

Hayes, M.H.B. (1997) Emerging concepts of the composition and structures of humic substances. In: Hayes, M.H.B. and Wilson, W.S. (eds) *Humic Substances in Soils, Peats and Waters*. The Royal Society of Chemistry, Cambridge, UK, pp. 1–30.

Hayes, M.H.B. and Swift, R.S. (1978) The chemistry of soil organic colloids. In: Gredenland, D.J. and Hayes, M.H.B. (eds) *Soil Colloids and their Associations in Aggregates*. Plenum, New York, pp. 179–320.

Hellämäki, M. (1992) Estimation of the nutritive value of silage using NIR digestibility. In: Murray, I. and Cowe, I.A. (eds) *Making Light Work: Advances in Near Infrared Spectroscopy*. VCH, Weinheim, Germany, pp. 264–268.

Henneberg, W. (1864) Analytisches Verfahren bei der Untersuchung der Futterstoffe, insbesondere der Stroh- und Heuarten. *Landwirtschaftlichen-Versuchs-Stationen* 6, 496–498.

Hildrum, K.I., Isaksson, T., Naes, T. and Tandberg, A. (1992) *Near Infra-red Spectroscopy*. Ellis Horwood, New York.

Hislop, J. and Cooke, I.J. (1968) Anion exchange resin as a means of assessing soil phosphate status: a laboratory technique. *Soil Science* 105(1), 8–11.

Hoermann, H. (1968) Colour reactions of carbohydrates. IV. Products of the colour reaction of D-fructose and D-glucose with anthrone-H_2SO_4. *Justus Liebig's Annalen der Chemie* 714, 174–190 (in German). See *Chemical Abstracts* 69, 59513x.

Horwitz, W. (2000) *Official Methods of Analysis of AOAC International*, 17th edn. AOAC International, Gaithersburg, Maryland, 2200+ pp.

Hughes, M. (1969) Determination of moisture and oil in the seed of winter rape (*Brassica napus*). Parts I & II. *Journal of the Science of Food and Agriculture* 20, 741–747.

Hyslop, J.J. and Cuddeford, D. (1996) Investigations on the use of the mobile bag technique in ponies. *Animal Science* 62(2), 647 (Abstract 99).

Hyslop, J.J., Stefansdottir, G.J., McLean, B.M.L., Longland, A.C. and Cuddeford, D. (1999) In situ incubation sequence and its effect on degradation of food components when measured in the caecum of ponies. *Animal Science* 69, 147–156.

Isaac, R.A. (1990) Plants. In: Helrich, K. (ed.) Official Methods of Analysis, 15th edn. *Association of Official Analytical Chemicals*, Arlington, Virginia.

Jacas, J., Marza, J., Florensa, P. and Soliva, M. (1987) Cation exchange capacity variation during the composting of different materials. In: De Bertoldi, M., Ferranti, M.P., L'Hermite, P. and Zucconi, F. (eds) *Compost: Production, Quality and Use*. Elsevier Applied Science, London, pp. 309–320.

Jackson, M.L. (1958) *Soil Chemical Analysis*. Prentice-Hall, Englewood Cliffs, New Jersey, pp. 175–176.

Jeffery, G.H., Bassett, J., Mendham, J. and Denney, R.C. (1989) *Vogel's Textbook of Quantitative Chemical Analysis* Longman, Harlow, UK.

Johnson, F.J. (1990a) Agricultural liming materials. In: Helrich, K. (ed.) *Official Methods of Analysis*, 15th edn. Association of Official Analytical Chemists, Arlington, Virginia, pp. 1–8.

Johnson, F.J. (1990b) Fertilizers. In: Helrich, K. (ed.) *Official Methods of Analysis*, 15th edn. Association of Official Analytical Chemists, Arlington, Virginia, p. 9–39.

Jones, D.I.H. and Hayward, M.V. (1973) A cellulase digestion technique for predicting the dry matter digestibility of forages. *Journal of the Science of Food and Agriculture* 24, 1419–1426.

Jones, D.I.H. and Hayward, M.V. (1975) The effect of pepsin pre-treatment of herbage on the prediction of dry matter digestibility from solubility in fungal cellulase solutions. *Journal of the Science of Food and Agriculture* 26, 711–718.

Jones, D.W. and Kay, J.J. (1976) Determination of volatile fatty acids C_1–C_6, and lactic acid in silage juice. *Journal of the Science of Food and Agriculture* 27, 1005–1014.

Kabata-Pendias, A. (2000) *Trace Elements in Soil and Plants*, 3rd edn. CRC Press, Boca Raton, Florida, 432 pp.

Kaneko, J.J. (1989) *Clinical Biochemistry of Domestic Animals*. Academic Press, San Diego, pp. 879–891.

Kaye, W. (1954) Near infrared spectroscopy I. Spectral identification and analytical applications. *Spectrochimica Acta* 6, 257–287.

Kjeldahl, J. (1883) Neue Methode zur Bestimmung des Stickstoffs in organischen Körpern. *Zeitschrift für Analytische Chemie* 22, 366–382.

Kotz, L., Kaiser, G., Tschöpel, P. and Tölg, G. (1972) Decomposition of biological materials for the determination of extremely low contents of trace elements in limited amounts with nitric acid under pressure in a Teflon tube. *Fresenius' Zeitschrift für Analytische Chemie* 260(3), 207–209.

Krischenko, V.P., Samokhvalov, L.G., Fomina,

L.G. and Novikova, G.A. (1992) Use of infrared spectroscopy for the determination of some properties of soil. In: Murray, I. and Cowe, I.A. (eds) *Making Light Work: Advances in Near Infrared Spectroscopy*. VCH, Weinheim, Germany, pp. 239–249.

Kubadinow, N. (1982) Zur Bestimmung von organischen Säuren in Preßschnitzelsilagen. *Zuckerindustrie* 107, 1107–1109.

Lachat Instruments (1988) QuikChem Method No. 12-115-01-1-A, Phosphorus as orthophosphate. *QuikChem Automated Ion Analyzer Methods Manual*. Milwaukee, Wisconsin.

Lampkin, N. and Measures, M. (eds) (1999) *1999 Organic Farm Management Handbook*, 3rd edn. University of Wales, Aberystwyth and Elm Farm Research Centre, Hamstead Marshall, UK, pp. 41–42.

Lessard, J.R., Briggs, R.A. and Scaletti, J.V. (1961) The organic acids in silage as determined by gas chromatography. *Canadian Journal of Plant Science* 41, 507–516.

Liebig, J. Von (1840) *Die Chemie in ihrer Anwendung auf Agrikultur und Physiologie*. [English editions: (1840) *Organic Chemistry in Its Applications to Agriculture and Physiology*. Edited from the manuscript of the author by Lyon Playfair, Taylor and Walton, London; and (1849) *Chemistry in Its Applications to Agriculture and Physiology*, Edited from the manuscript of the author by Lyon Playfair and William Gregory, 4th edn. revised and enlarged, John Wiley & Sons, New York].

Little, T.M. and Hills, F.J. (1978) *Agricultural Experimentation*, John Wiley and Sons, New York, 350 pp.

Longland, A.C. (2001) Plant carbohydrates: analytical methods and nutritional implications for equines. Internal communication, Institute of Grassland and Environmental Research, Aberystwyth, UK.

Luginbuhl, J.M., Pond, K.R. and Burns, J.C. (1994) Whole-tract digesta kinetics and comparison techniques for the estimation of fecal output in steers fed coastal bermudagrass hay at four levels of intake. *Journal of Animal Science* 75, 201–211.

Lund, H.A., Swanson, C.L.W. and Jacobson, H.G.M. (1950) The Morgan soil testing system. *Connecticut Agricultural Experiment Station Bulletin* No. 541.

Luxon, S.G. (ed.) (1992) *Hazards in the Chemical Laboratory*, 5th edn. Royal Society of Chemistry, Cambridge, 676 pp.

Machelboeuf, D., Marangi, M., Poncet, C. and Martin-Rosset, W. (1995) Study of nitrogen digestion from different hays by the mobile bag technique in horses. *Annales de Zootechnie* 44, Supplement 219.

Machelboeuf, D., Poncet, C., Jestin, M. and Martin-Rosset, W. (1996) Use of a mobile nylon bag technique with caecum fistulated horses as an alternative method for estimating precaecal and total tract nitrogen digestibility. *Proceedings of the 47th Annual Meeting of the European Association for Animal Production*, Lillehammer, Norway, p. 269.

Mader, T.L., Teeter, R.G. and Horn, G.W. (1984) Comparison of forage labelling techniques for conducting passage rate studies. *Journal of Animal Science* 58(1), 208–212.

MAFF (1990) *UK Tables of Nutritive Value and Chemical Composition of Feedingstuffs*, Givens, D.I. and Moss, A.R. (eds) MAFF Standing Committee on Tables of Feed Composition, Rowett Research Services, Aberdeen, 420 pp.

MAFF (1993a) Appendix I. The determination of oil in feeding stuffs (Based on SI 1985 No. 1119). In: *Prediction of the Energy Values of Compound Feeding Stuffs for Farm Animals*. MAFF, PB 1285, 9–11.

MAFF (1993b) Appendix II. Determination of neutral cellulase plus gamanase digestibility (NCGD) of feeding stuffs. In: *Prediction of the Energy Values of Compound Feeding Stuffs for Farm Animals*. MAFF, PB 1285, 12–15.

MAFF (1993c) Appendix III. The determination of neutral detergent (plus amylase) fibre (NDF) of feeding stuffs. In:

Prediction of the Energy Values of Compound Feeding Stuffs for Farm Animals. MAFF, PB 1285, 16–17.

MAFF/ADAS (1971) *Nutrient Allowances and Composition of Feedingstuffs for Ruminants*. ADAS Advisory Paper No. 1. MAFF, Pinner, UK, 17 pp.

MAFF/ADAS (1986) *The Analysis of Agricultural Materials*, Reference Book 427. HMSO, London.

MAFF/ADAS (2000) *Fertiliser Recommendations*, 7th edn. Reference Book 209, The Stationery Office, London, p. 167.

Mahler, R.L., Naylor, D.V. and Frederickson, M.K. (1984) Hot water extraction of boron from soils using sealed plastic pouches. *Communications in Soil Science and Plant Analysis* 15, 479–492.

Malley, D.F. and Nilsson, M. (1995) Environmental applications of near infrared spectroscopy: seeing the environment in a different light. *Spectroscopy Europe* 7(6), 8–16.

Marais, J.P. (2000) Use of markers. In: D'Mello, J.P.F. (ed.) *Farm Animal Metabolism and Nutrition*. CAB International, Wallingford, UK, pp. 255–277.

Mathur, S.P., Owen, G., Dinel, H. and Schnitzer, M. (1993) Determination of Compost Biomaturity. I. Literature Review. *Biological Agriculture and Horticulture* 10, 65–85. [Viewable at http://www.nes.coventry.ac.uk/bah/classic.htm.]

McDonald, P., Edwards, R.A. and Greenhalgh, J.F.D. (1969) *Animal Nutrition*. Oliver and Boyd, Edinburgh, pp. 121–122.

McDonald, P., Edwards, R.A., Greenhalgh, J.F.D. and Morgan, C.A. (1995) *Animal Nutrition*, 5th edn. Longman Scientific and Technical, Harlow, 607 pp.

McGrath, D. (1997) Extractability, chemical composition, and reactivities of soil organic matter of Irish grassland soils. In: Hayes, M.H.B. and Wilson, W.S. (eds) *Humic Substances in Soils, Peats and Waters*. The Royal Society of Chemistry, Cambridge, UK, pp. 31–38.

McIntosh, J.L. (1969) Bray and Morgan soil test extractants modified for testing acid soils from different parent materials. *Agronomy Journal* 61, 259–265.

McLean, E.O. (1977) Contrasting concepts in soil test interpretation: sufficiency levels of available nutrients versus basic cation saturation ratios. Soil testing: correlating and interpreting the analytical results. *ASA Special Publication* No. 29, 39–54, American Society of Agronomy, Madison, Wisconsin.

McLean, E.O., Hartwig, R.C., Eckert, D.J. and Triplett, G.B. (1983) Base cation saturation ratios as a basis for fertilizing and liming: II. Field studies. *Agronomy Journal* 75, 635–639.

Mead, R., Curnow, R.N. and Hasted, A.M. (1993) *Statistical Methods in Agriculture and Biology*, 2nd edn. Chapman & Hall, New York, 415 pp.

Mehlich, A. (1953) Determination of P, Ca, Mg, F, Na and NH_4. *North Carolina Soil Test Division* (Mimeo).

Mehlich, A. (1984) Mehlich 3 soil test extractant: a modification of the Mehlich 2 extractant. *Commununications in Soil Science and Plant Analysis* 15(12), 1409–1416.

Meier, P.C. and Zünd, R.E. (1993) *Statistical Methods in Analytical Chemistry*. John Wiley & Sons, New York, 321 pp.

Mercier, P. (ed.) (1996) *Laboratory Safety Pocket Handbook*. Genium Publishing Corporation, Amsterdam, New York.

Miall, S. and Miall, L.M. (1956) *A New Dictionary of Chemistry*. Longmans, Green, London, p. 101.

Middleton, G. and Stuckey, R.E. (1953) The preparation of biological material for the determination of trace metals. Part I. A critical review of existing procedures. *Analyst* 78, 532–542.

Middleton, G. and Stuckey, R.E. (1954) The preparation of biological material for the determination of trace metals. Part II. A method for the destruction of organic matter in biological material. *Analyst* 79, 138–142.

Milham, P.J., Awad, A.S., Paull, R.E. and Bull, J.H. (1970) Analysis of plants, soils and waters for nitrate by using an ion-selective electrode. *Analyst* 95, 751–757.

Missouri Agricultural Experiment Station

(1998) *Recommended Chemical Soil Test Procedures for the North Central Region.* North Central Region Research Publication No. 221 (Revised). To order, request SB1001, *Recommended Chemical Soil Test Procedures for the North Central Region* ($6). Telephone orders: +1 573 882–7216 or (1–800–292–0969 within USA).

Moir, K.W. (1982) Theory and practice of measuring the cell-wall content of food for ruminant and non-ruminant animals. *Laboratory Practice* 31, 732–733. Also (1972) *Journal of the Agricultural Society, Cambridge* 78, 351–353.

Moore-Colyer, M.J.S. and Longland, A.C. (2000) Intakes and *in vivo* apparent digestibilities of four types of conserved grass forage by ponies. *Animal Science* 71, 527–534.

Moore-Colyer, M.J.S., Hyslop, J.J., Longland, A.C. and Cuddeford, D. (1997a) Degradation of four dietary fibre sources by ponies as measured by the mobile bag technique. In: *Proceedings of the 15th Equine Nutrition and Physiology Symposium.* Fort Worth, Texas, pp. 118–119.

Moore-Colyer, M.J.S., Hyslop, J.J., Longland, A.C. and Cuddeford, D. (1997b) The degradation of organic matter and crude protein of four botanically diverse feedstuffs in the foregut of ponies as measured by the mobile bag technique. In: *Proceedings of the British Society of Animal Science*, Scarborough, p. 120.

Morgan, N.F. (1941) Chemical soil diagnosis by the universal soil testing system. *Connecticut Agricultural Experiment Station Bulletin* No. 450.

Morrow, H.J. (1998) Fermentation kinetics and *in vivo* apparent digestibilities and rates of passage of two chop lengths of big bale silage and hay in ponies. MSc. thesis, University of Wales, Aberystwyth, 131 pp.

Moss, A.R. and Givens, D.I. (1990) Chemical composition and *in vitro* digestion to predict digestibility of field-cured and barn-dried grass hays. *Animal Feed Science Technology* 31, 125–138.

Murray, I. and Williams, P.C. (1987) In: Williams, P. and Norris, K. (eds) *Near-Infrared Technology in the Agricultural and Food Industries.* American Association of Cereal Chemists, St Paul, Minnesota, p. 22.

Nocek, J.E. (1988) *In situ,* and other methods to estimate ruminal protein and energy digestibility. A review. *Journal of Dairy Science* 71, 2051–2069.

Nozière, P. and Michalet-Doreau, B. (2000) *In sacco* methods. In: D'Mello, J.P.F. (ed.) *Farm Animal Metabolism and Nutrition.* CAB International, Wallingford, UK, pp. 233–253.

Nyberg, M.A., Potter, G.D., Gibbs, P.G., Schumacher, J., Murray-Gerzik, M., Bombarda, A. and Swinney, D.L. (1995) Flow rate through the equine small intestine determined with soluble and insoluble indicators. *Equine Nutrition and Physiology Symposium* No. 14, 36–41.

OEEC (1952) *Fertilisers – Methods of Analysis used in OEEC countries.* Organisation for European Economic Co-operation, Paris, 182 pp.

Offer, N.W. (1993) *Proceedings of the Society of Feed Technologists Ruminants Conference, Coventry.* Published by Society of Feed Technologists, Reading, pp. E1–E7.

Olsen, S.R., Cole, C.V., Watanabe, F.S. and Dean, L.A. (1954) Estimation of available phosphorus in soils by extraction with sodium bicarbonate. *USDA Circular No. 939*, US Government Printing Office, Washington, DC.

Olsen, S.R. and Dean, L.A. (1965) Phosphorus. In: Black, C.A. (ed.) *Methods of Soil Analysis, Part 2.* American Society of Agronomy, Madison, Wisconsin, pp. 1035–1049.

Olson, R.A., Rhodes, M.B. and Dreier, A.F. (1954) Available phosphorus status in Nebraska soils in relation to series classification, time of sampling and method of measurement. *Agronomy Journal* 46, 175–180.

Omed, H.M., Lovett, D.K. and Axford, R.F.E. (2000) Faeces as a source of microbial enzymes for estimating digestibility. In:

Forage Evaluation in Ruminant Nutrition. CAB International, Wallingford, UK, pp. 135–154.

Ørskov, E.R. and McDonald, I. (1979) The estimation of protein degradability in the rumen from incubation measurements weighted according to rate of passage. *Journal of Agricultural Science (Cambridge)* 92, 499–503.

Osborne, B.G. and Fearn, T. (1986) *Near Infrared Spectroscopy in Food Analysis*, Longman Scientific and Technical, UK and John Wiley & Sons, New York.

Padmore, J.M. (1990) Animal feed, In: Helrich, K. (ed.) *Official Methods of Analysis*, 15th edn. Association of Official Analytical Chemists, Arlington, Virginia.

Parnell, A. and White, J. (1983) The use of near infra-red reflectance analysis in predicting the digestibility and the protein and water soluble carbohydrate contents of perennial ryegrass. *Journal of the National Institute of Agricultural Botany* 16, 221–227.

Paul, J.L. and Carlson, R.M. (1968) Nitrate determination in plant extracts by the nitrate electrode. *Journal of Agricultural and Food Chemistry* 16, 766–768.

PDA (1999a) *Potash and Biosolids.* Technical Leaflet No. 20. The Potash Development Association, Laugharne, Carmarthen, UK, 12 pp.

PDA (1999b) *Potash for Organic Growers.* Technical Leaflet No. 23. The Potash Development Association, Laugharne, Carmarthen, UK, 12 pp.

PDA (1999c) *Effective Use of Soil Analysis.* Technical Leaflet No. 24. The Potash Development Association, Laugharne, Carmarthen, UK, 12 pp.

PDA (2000) Development in soil analysis. *Potash News*, June 2000, The Potash Development Association, Laugharne, Carmarthen, UK, 2 pp.

PDA (2001) *Potash for Cereals.* Technical Leaflet No. 11. The Potash Development Association, Laugharne, Carmarthen, UK, 16 pp.

Pearson, R.A. and Merritt, J.B. (1991) Intake, digestion and gastrointestinal transit time in resting donkeys and ponies and exercised donkeys given *ad libitum* hay and straw diets. *Equine Veterinary Journal* 23(5), 339–343.

Perkins, H.F. (1970) A rapid method of evaluating the zinc status of coastal plain soils. *Communications in Soil Science and Plant Analysis* 1, 35–42.

Perry, T.W. (1980) *Beef Cattle Feeding and Nutrition.* Academic Press, New York, pp. 364–368.

Playne, M.J. (1985) Determination of ethanol, volatile fatty acids, lactic and succinic acids in fermentation liquids by gas chromatography. *Journal of the Science of Food and Agriculture* 36, 638–644.

Poppi, D.P., France, J. and McLennan, S.R. (2000) Intake, passage and digestibility. In: Theodorou, M.K. and France, J. (eds) *Feeding Systems and Feed Evaluation Models.* CAB International, Wallingford, UK, pp. 35–52.

Pratt, P.F. (1965) Potassium. In: *Methods of Soil Analysis, Part 2*, 2nd edn. American Society of Agronomy, p. 1022. (1994) Cornell University Press, Ithaca, New York.

Reeves, J.B. III and McCarty, G.W. (2001) Quantitative analysis of agricultural soils using near infrared reflectance spectroscopy and a fibre-optic probe. *Journal of Near Infrared Spectroscopy* 9, 25–34.

Reeves, J.B. III and Van Kessel, J.S. (2000) Determination of ammonium-N, moisture, total C and total N in dairy manures using a near infrared fibre-optic spectrometer. *Journal of Near Infrared Spectroscopy* 8, 151–160.

Reeves, J.B. III, McCarty, G.W. and Meisinger, J.J. (2000) Near infrared reflectance spectroscopy for the determination of biological activity in agricultural soils. *Journal of Near Infrared Spectroscopy* 8, 161–170.

Reisenauer, H.M. (1965) Molybdenum. In: Black, C.A. (ed.) *Methods of Soil Analysis.* American Society of Agronomy, Madison, Wisconsin, pp. 1050–1058.

Rooke, J.A., Borman, A.J. and Armstrong, D.G. (1990) The effect of inoculation with *Lactobacillus plantarum* on fermentation in laboratory silos of herbage low in water soluble carbohydrate. *Grass and Forage Science* 45, 143–152.

Ross, P.F. (1990) Veterinary analytical toxicology. In: *Official Methods of Analysis*, 15th edn. AOAC, Arlington, p. 356.

Rowell, D.L. (1994) *Soil Science: Methods and Applications*. Longman, Harlow, UK.

[RSC] Royal Society of Chemistry (1996) *COSHH in Laboratories*, 2nd edn. Royal Society of Chemistry, Cambridge, 34 pp.

Ruzicka, J. and Hansen, E.H. (1975) Flow injection analysis. Part 1. A new concept of fast continuous flow analysis. *Analytica Chimica Acta* 78, 145–157.

Salawu, M.B., Acamovic, T., Stewart, C.S., Hovell, F.D.DeB. and McKay, I. (1997) Assessment of the nutritive value of *Calliandra calothyrsus*: *in sacco* degradation and *in vitro* gas production in the presence of Quebracho tannins with or without Browse Plus. *Animal Feed Science Technology* 69, 219–232.

Sauchelli, V. (1965) *Phosphates in Agriculture*. Reinhold Publishing, New York, 277 pp.

Schlichting, E. and Blume, H.P. (1961) *Bodenkundliches Praktikum*. Verlag Paul Parey, Berlin, 209 pp.

Schmid, O. (1984) Chemical soil analysis methods in biological husbandry. In: Lampkin, N. and Woodward, L. (eds) *The Soil: Assessment, Analysis and Utilisation in Organic Agriculture*. Elm Farm Research Centre, Practical Handbook Series, EFRC, Hamstead Marshall, UK, pp. 36–43.

Schneider, B.H. and Flatt, W.P. (1975) *The Evaluation of Feeds through Digestibility Experiments*. University of Georgia Press, Athens, 423 pp.

Schnitzer, M. and Khan, S.U. (1972) *Humic Substances in the Environment*. Marcel Dekker, New York, 327 pp.

Schofield, R.K and Taylor, A.W. (1955) The measurement of soil pH. *Soil Science Society of America Proceedings*. 19, 164–167.

Schwer, E.W. and Conan, H.R. (1960) Fertiliser analysis. The determination of potassium. In: *Joint Symposium on Fertiliser Analysis. Proceedings No. 62*.

The Fertiliser Society, London, pp. 99–125; Discussion, pp. 126–136.

Scott, R.O., Mitchell, R.L., Purves, D. and Voss, R.C. (1971) Spectrochemical methods for the analysis of soils, plants and other agricultural materials. *Bulletin 2*. Macaulay Institute for Soil Research, Aberdeen, p. 8.

Shenk, J.S., Workman, J.J. Jr and Westerhaus, M.O. (1992) Application of NIR spectroscopy to agricultural products. In: *Handbook of Near-Infrared Analysis*. Marcel Dekker, New York, pp. 383–429.

Siegfried, Von R., Rückemann, H. and Stumpf, G. (1984) Eine HPLC-Methode zur Bestimmung organischer Säuren in Silagen. *Landwirtschaftliche Forschung* 37, 3–4.

Simpson, A.J., Watt, B.E., Graham, C.L. and Hayes, M.H.B. (1997) Humic substances from podzols under oak forest and a cleared forest site I. Isolation and characterization. In: Hayes, M.H.B. and Wilson, W.S. (eds) *Humic Substances in Soils, Peats and Waters*. The Royal Society of Chemistry, Cambridge, UK, pp. 73–82.

Skeggs, L.T. Jr (1957) Automatic method for colorimetric analysis. *American Journal of Clinical Pathology* 28, 311–322.

Slavin, W. (1968) *Atomic Absorption Spectroscopy*. John Wiley Interscience, New York, 307 pp., p. 73.

Solangi, A.A. (1997) Studies on the use of faecal organisms in the *in vitro* assessment of forages. Ph.D. thesis, University of Wales, Bangor, UK.

Somasire, L.L.W. and Edwards, A.C. (1992) An ion exchange resin method for nutrient extraction of agricultural advisory soil samples. *Communications in Soil Science and Plant Analysis* 23(7&8), 645–657.

Southgate, D.A.T. (1995) *Dietary Fibre Analysis*. Royal Society of Chemistry, Cambridge, UK.

Stahr, H.M. (1991) *Analytical Methods in Toxicology*. John Wiley & Sons, New York, p. 322.

Stakelum, G., Morgan, D. and Dillon, P. (1988) A comparison of *in vitro* procedures for estimating herbage digestibility.

Irish Journal of Agricultural Research 27, 104–105.

Stark, S and Luchter, K. (1991) Overview of near infrared spectroscopy. *Fourth International Conference on Near Infrared Spectroscopy*, Aberdeen, 18–23 August, 1991.

Statutory Instrument No. 1342 (1996) The Fertilisers (Sampling and Analysis) Regulations 1996. The Stationery Office, London (web version). The title page of the SI may be downloaded from the following website: http://www.hmso.gov.uk/si/si1996/Uksi_19961342_en_1.htm with Schedule 2, *Methods of Analysis* at: http://www.hmso.gov.uk/si/si1996/Uksi_19961342_en_4.htm#sdiv2.

Stockdale, E.A. (2001) Optimisation of phosphorus and potassium management within organic farming systems. *Review Report No. OF0114*, from Review of DEFRA's R & D on Organic Farming, 12–13 July 2001, Warwick. (DEFRA = Department for Environment, Food and Rural Affairs; see http://www.defra.gov.uk/research/researchfrm.htm)

Stockdale, E.A., Lampkin, N.H., Hovi, M., Keatinge, R., Lennartsson, E.K.M., Macdonald, D.W., Padel, S., Tattersall, F.H., Wolfe, M.S. and Watson, C.A. (2000) Agronomic and environmental implications of organic farming systems. *Advances in Agronomy* 70, 261–327.

Suehara, K., Nakano, Y. and Yano, T. (2001) Simultaneous measurement of carbon and nitrogen content of compost using near infrared spectroscopy. *Journal of Near Infrared Spectroscopy* 9, 35–41.

Sun, L., Gao, Z., Li, L., Yu, X. and Fang, Z. (1981) Determination of soil available phosphorus by flow injection analysis [in Chinese, English translation]. *Fenxi Huaxue (Analytical Chemistry)* 9, 586.

Suzuki, M. and Lund, C.W. (1980) Improved gas–liquid chromatography for simultaneous determination of volatile fatty acids and lactic acid in silage. *Journal of Agricultural and Food Chemistry* 28, 1040–1041.

Swift, R.S. (1996) Organic matter characterization. In: Sparks, D.L., Page, A.L., Helmke, P.A., Loeppert, R.H., Soltanpour, P.N., Tabatabai, M.A., Johnson, C.T. and Sumner, M.E. (eds) *Methods of Soil Analysis. Part 3. Chemical methods*. SSSA Book Series No. 5, Soil Science Society of America and American Society of Agronomy, Madison, Wisconsin, pp. 1011–1069.

[TCA] The Composting Association (2000) Standards for composts. *Briefing Note May 2000*. The Composting Association, p. 4.

Teeter, R.G., Owens, F.N. and Horn, G.W. (1979) Ytterbium as a ruminal marker. *Journal of Animal Science* 49, Supplement 1, 412.

Thomas, M.J.K. (1996) *Ultraviolet and Visible Spectroscopy*, 2nd edn. ACOL Series, John Wiley & Sons, Chichester, UK, 229 pp.

Thomas, T.A. (1977) An automated procedure for the determination of soluble carbohydrates in herbage. *Journal of the Science of Food and Agriculture* 28, 639–642.

Tilley, J.M.A. and Terry, R.A. (1963) A two-stage technique for the *in vitro* digestion of forage crops. *Journal of the British Grassland Society* 18, 104–111.

Tinsley, J. (1950) Determination of organic carbon in soils by dichromate mixtures. *Proceedings of the 4th International Congress of Soil Science*, Vol. 1, 161.

Tomlinson, A.L. (1997) Voluntary food intake and apparent digestibility of grass chaff, and an assessment of mobile bag technique to study the dynamics of fibre digestion in equids. M.Sc. thesis, University of Wales, Aberystwyth, UK, 118 pp.

Triebold, H.O. (1946) *Quantitative Analysis with Applications to Agricultural and Food Products*. Van Nostrand, New York, pp. 18–19.

Truog, E. (1930) Determination of the readily available phosphorus of soils. *Journal of the American Society of Agronomy* 22, 874–882.

Truog, E. and Meyer, A.H. (1929) Improvements in the Denigès colorimetric method for phosphorus and arsenic. *Industrial and Engineering Chemistry, Analytical Edition*, 1, 136–139.

Turner, B.L. and Haygarth, P.M. (2000) Phosphorus forms and concentrations in leachate under four grassland soil types. *Soil Science Society of America Journal* 64(3), 1090–1099.

USDA (1996) *Soil Survey Laboratory Methods Manual*, Burt, R. (ed.). Soil Survey Investigations Report No. 42, Version 3.0, United States Department of Agriculture, Washington, DC, 693 pp.

Van Handel, E. (1967) Determination of fructose and fructose yielding carbohydrates with cold anthrone. *Analytical Biochemistry* 19(1), 193–194.

Van Soest, P.J. (1973) Collaborative study of acid-detergent fibre and lignin. *Journal of the Association of Official Analytical Chemists* 56, 781–784.

Van Soest, P.J. (1982) *Nutritional Ecology of the Ruminant.* O & B Books, Corvallis, Oregon, 374 pp. (2nd edn. (1994), Ithaca, London, 476 pp.).

Van Soest, P.J. and Wine, R.H. (1967) Use of detergents in the analysis of fibrous feeds. IV. Determination of plant cell-wall constituents. *Journal of the Association of Official Analytical Chemists* 50(1), 50–55.

Van Soest, P.J. and Wine, R.H. (1968) Determination of lignin and cellulose in ADF with permanganate. *Journal of the Association of Official Analytical Chemists* 51(4), 780–785.

Van Soest, P.J., Wine, R.H. and Moore, L.A. (1966) Estimation of the true digestibility of forages by the *in vitro* digestion of cell walls. *Proceedings of the Xth International Grassland Congress,* Helsinki, pp. 438–441.

Van Soest, P.J., Robertson, J.B. and Lewis, B.A. (1991) Methods for dietary fiber, neutral detergent fiber, and nonstarch polysaccharides in relation to animal nutrition. *Journal of Dairy Science* 74, 3583–3597.

VDLUFA (1991) *Die Untersuchung von Böden, Methodenbuch Nr. 1,* VDLUFA-Verlag, Darmstadt, Germany, 970 pp. [VDLUFA = Verband Deutscher Landwirtschaftlicher Untersuchungs- und Forschungsanstalten].

VDLUFA (1995) *Die Untersuchung von Düngmitteln, Methodenbuch Nr. II,* VDL-UFA-Verlag, Darmstadt, Germany, Section 4.1.3.

Walkley, A. and Black, I.A. (1934) An examination of Degtjareff method for determining soil organic matter and a proposed modification of the chromic acid titration method. *Soil Science* 37, 29–37.

Wallinga, I., van der Lee, J.J., Houba, V.J.G., van Vark, W. and Novozamsky, I. (eds) (1995) *Plant Analysis Manual.* Kluwer Academic Publishers, Dordrecht, Netherlands.

Walters, C. Jr (1975) (ed.) *The Albrecht Papers,* Vol. II. *Soil Fertility and Animal Health.* Acres USA, Kansas City, Missouri, 192 pp.

Walters, C. Jr (1989) (ed.) *The Albrecht Papers.* Vol. III. *Hidden Lessons in Unopened Books.* Acres USA, Kansas City, Missouri, 401 pp.

Walters, C. Jr (1992) (ed.) *The Albrecht Papers,* Vol. IV, *Enter Without Knocking.* Acres USA, Kansas City, Missouri, 315 pp.

Walters, C. Jr (1996) (ed.) *The Albrecht Papers,* Vol. I, *Foundation Concepts.* Acres USA, Metairie, Louisiana, 515 pp.

Watson, C. (1994) *Official and Standardized Methods of Analysis,* 3rd edn. Royal Society of Chemistry, Cambridge, UK, 778 pp.

Wehmer, C. (1929) *Die Pflanzenstoffe,* Vol. 1. Verlag von Gustav Fischer, Jena, Germany, pp. 1–640.

Wehmer, C. (1931) *Die Pflanzenstoffe,* Vol. 2. Verlag von Gustav Fischer, Jena, Germany, pp. 641–1511.

Welte, E. (1955) Neurere Ergebnisse der Humusforschung. *Angewandte Chemie* 67(5), 153–155.

Whitehead, D.C. (1966) *Data on the Mineral Composition of Grassland Herbage.* Technical Report No. 4, Grassland Research Institute, Hurley, UK, 55 pp.

Williams, C.H. (1950) Studies on soil phosphorus. I. A method for the partial fractionation of soil phosphorus. *Journal of Agricultural Science* 40, 233–242.

Williams, P.C. (1991) Application of near-infrared reflectance and transmittance spectroscopy in agriculture. In: Biston, R. and Bartiaux-Thill, N. (eds) *Proceedings of the Third International Conference on Near Infrared Spectroscopy*, Brussels, Vol. 2, pp. 463–473.

Williams, P. and Norris, K. (1987) *Near-Infrared Technology in the Agricultural and Food Industries.* American Association of Cereal Chemists, St Paul, Minnesota, 330 pp. [Note: Second edition was printed in 1990.]

Wilman, D. and Altimimi, M.A.K. (1982) The digestibility and chemical composition of plant parts in Italian and perennial rye-grass during primary growth. *Journal of the Science of Food and Agriculture*, 33, 595–602.

Wilman, D. and Wright, P.T. (1978) Dry-matter content, leaf water potential and digestibility of three grasses in the early stages of regrowth after defoliation with and without applied nitrogen. *Journal of Agricultural Science, Cambridge* 91, 365–380.

Woolford, M.K. (1984) *The Silage Fermentation.* Marcel Dekker, New York, 139 pp.

Yemm, E.W. and Willis, A.J. (1954) The estimation of carbohydrates in plant extracts by anthrone. *Biochemistry Journal* 57, 508–514.

Zucconi, F. and De Bertoldi, M. (1987) Compost specifications for the production and characterization of compost from municipal solid waste. In: De Bertoldi, M., Ferranti, M.P., L'Hermite, P. and Zucconi, F. (eds) *Compost: Production, Quality and Use.* Elsevier Applied Science, London, pp. 30–50.

Appendix 1
Equipment Suppliers

Used Equipment Suppliers

UK

Sarose Scientific Instruments:	http://www.sarose.3av.com
	e-mail: sarose@compuserve.com
Science Exchange Service:	http://www.science-exchange.com
Severn Sales:	Olveston Road, Horfield, Bristol BS7 9PB, UK. Tel. +44 (0) 0117 9354125
Severn Science:	Short Way, Thornbury, Bristol BS12 2UL
	Tel. +44 (0) 1454 414723; fax +44 (0) 1454 417101
Spectro-Service Ltd:	Tel. 01280 705577; fax 01280 705510
Tecmec Services:	http://www.atomicabsorption.co.uk

USA

Analytical Instrument Recycle, Inc.:	http://www.aironline.com
Encore Lab & Analytical:	http://www.encorelab.com
GenTech:	http://www.gcmsservice.com/equip_home.htm
IET Ltd:	http://www.ietltd.com

Canada

GSR Technical Sales:	http://gsrtech.com
Labequip Ltd:	http://www.labequip.org

General

LabX (dealer links)	http://www.labx.com

Internet auctions

SciQuest:	http://www.auctions-sciquest.com/

New Equipment Suppliers

Web addresses of individual suppliers are given in the text, but a general search facility of over 1000 worldwide manufacturers is available at:
http://www.product-search.co.uk

Appendix 2
Soil Index Table

Index	Magnesium (mg l^{-1})	Phosphorus Olsen's P (mg l^{-1})	Potassium (mg l^{-1})
0	0–25	0–9	0–60
1	26–50	10–15	61–120
2	51–100	16–25	121–180 (2–) 181–240 (2+)
3	101–175	26–45	241–400
4	176–250	46–70	401–600
5	251–350	71–100	601–900
6	351–600	101–140	901–1500
7	601–1000	141–200	1501–2400
8	1001–1500	201–280	2401–3600
9	>1500	>280	>3600

ADAS Classification, MAFF/ADAS, 2000.

Appendix 3
Lime Application Rates for Arable Land

Values are in tons per acre (t ha^{-1} in parentheses) ground/magnesium lime-stone or chalk of neutralizing value 50–55, with fineness of 40% passing through a 150 μm (No. 100) sieve, and to bring the top 20 cm (8 in) of soil to the optimum pH of 6.5 (5.8 for peaty soil).

pH	Sandy	Loam	Clay	Peaty
6.2	1.25(3.0)	1.5(4.0)	1.5(4.0)	0
6.1	1.5(4.0)	1.5(4.0)	2.0(5.0)	0
6.0	1.5(4.0)	2.0(5.0)	2.5(6.0)	0
5.9	2.0(5.0)	2.5(6.0)	2.5(6.0)	0
5.8	2.0(5.0)	2.5(6.0)	3.0(7.0)	0
5.7	2.5(6.0)	3.0(7.0)	3.5(8.0)	1.0(2.5)
5.6	3.0(7.0)	3.5(8.0)	3.75(9.0)	2.0(5.0)
5.5	3.0(7.0)	3.5(8.0)	4.0(10.0)	3.5(8.0)
5.4	3.5(8.0)	3.75(9.0)	4.0(10.0)	4.0(10.0)
5.3	3.5(8.0)	4.0(10.0)	4.5(11.0)	4.5(11.0)
5.2	3.75(9.0)	4.5(11.0)	5.0(12.0)	5.25(13.0)
5.1	4(10.0)	4.5(11.0)	5.25(13.0)	6.0(14.0)
5.0	4(10.0)	5.0(12.0)	6.0(14.0)	6.5(16.0)

Based on MAFF/ADAS (2000) and other material.

Appendix 4
Lime Application Rates for Grassland

Values are in tons per acre (t ha^{-1} in parentheses) ground/magnesium limestone or chalk of neutralizing value 50–55, with fineness of 40% passing through a 150 μm (No. 100) sieve, and to bring the top 15 cm (6 in) of soil to the optimum pH of 6.0 (5.3 for peaty soil).

pH	Sandy	Loam	Clay	Peaty
6.0	0	0	0	0
5.9	0.75(2.0)	0.75(2.0)	0.75(2.0)	0
5.8	0.75(2.0)	0.75(2.0)	0.75(2.0)	0
5.7	0.75(2.0)	1.25(3.0)	1.25(3.0)	0
5.6	1.25(3.0)	1.25(3.0)	1.5(4.0)	0
5.5	1.25(3.0)	1.5(4.0)	1.5(4.0)	0
5.4	1.5(4.0)	1.5(4.0)	2.0(5.0)	0
5.3	1.5(4.0)	2.0(5.0)	2.0(5.0)	0
5.2	2.0(5.0)	2.0(5.0)	2.5(6.0)	1.5(4.0)
5.1	2.0(5.0)	2.5(6.0)	3.0(7.0)	2.0(5.0)
5.0	2.0(5.0)	2.5(6.0)	3.0(7.0)	2.5(6.0)
4.9	2.5(6.0)	3.0(7.0)	3.0(7.0)	3.0(7.0)
4.8	2.5(6.0)	3.0(7.0)	3.0(7.0)	3.0(7.0)

Based on MAFF/ADAS (2000) and other material.

Appendix 5
Nitrate and Nitrite in Soil, Plant and Fertilizer Extracts[a]

Ranges: 0–3 to 0–25 mg l^{-1} as NO_3; 0–1.2 to 0–5.64 mg l^{-1} as N

Description

This automated procedure for the determination of nitrate and nitrite uses the procedure whereby nitrate is reduced to nitrite by a copper–cadmium reductor column.[1,2] The nitrite ion then reacts with sulphanilamide under acidic conditions to form a diazo compound. This compound then couples with N-1-naphthylethylenediamine dihydrochloride to form a reddish-purple azo dye.

Hardware: Cd-reductor **Pump tubes:** 4 + 2 air + 1 sampler wash (AAII: +1)

Typical performance data

Test conditions: range: 0–5.6 mg N l^{-1}

Sampling rate	40 h^{-1}
Sample: wash ratio	5:1
Reagent absorbance	0.01
Sensitivity: extinction at 5.6 mg l^{-1} as N	0.29–0.33

[a]Method No. G-067-X (Copyright Bran+Luebbe. Not to be reproduced without permission).

Coefficient of variation: (replicates at 50%)	0.39%
Pooled standard deviation: (25 at 5 levels)	0.007 mg l^{-1} as N
Correlation coefficient: (linear fit)	0.9999
Detection limit (determined according to EPA procedure pt. 136, app. B)	0.007 mg l^{-1} as N

Note: these performance specifications were developed with the exclusive use of genuine Bran+Luebbe parts and consumables.

References

1. Armstrong, F.A.J., Sterns, C.R. and Strickland, J.D.H. (1967) The measurement of upwelling and subsequent biological processes by means of the Technicon AutoAnalyzer and associated equipment. *Deep-Sea Research* 14, 381–389.
2. Grasshoff, K. (1969) Technicon International Congress, June 1969.

Reagents

Unless otherwise specified all chemicals should be of ACS grade or equivalent.

List of raw materials

	Safety classification
Ammonium chloride, NH_4Cl	harmful
Ammonium hydroxide, NH_4OH, conc.	irritant
Brij-35*, 30% solution (Bran+Luebbe No. T21-0110-06)	–
Cadmium, powder (Bran+Luebbe part no. T11-5063)	toxic
Chloroform, $CHCl_3$	harmful
Copper (II) sulphate pentahydrate, $CuSO_4.5H_2O$	harmful
Hydrochloric acid, 37%, HCl	corrosive
N-(1–Naphthyl)ethylene diamine dihydrochloride, $C_{12}H_{14}N_2.2HCl$	irritant
Phosphoric acid, 85%, H_3PO_4	corrosive
Potassium chloride, KCl	–
Potassium nitrate nonahydrate, $KNO_3.9H_2O$	oxidizing, irritating
Potassium nitrate (KNO_3)	oxidizing
Sulphanilamide, $C_6H_8N_2O_2S$	–

Reagent make-up

DI water refers to high quality reagent water, Type I or Type II as defined in ASTM Standards, Part 31, D 1193-74.

---

* Registered Trademark of Atlas Chemical Industries Inc.

Ammonium chloride reagent

Ammonium chloride	10 g
Water	to 1000 ml
Brij-35, 30% solution	0.5 ml

Dissolve 10 g of ammonium chloride in water and dilute to 1000 ml. Add 0.5 ml of Brij-35 and mix thoroughly. Adjust the pH to 8.5 with ammonia.

Colour reagent

Sulphanilamide	10 g
Phosphoric acid, conc.	100 ml
N-1-Naphthylethylenediamine dihydrochloride	0.5 g
DI water	to 1000 ml
Brij-35, 30% solution	0.5 ml

To approximately 700 ml of DI water add 100 ml concentrated phosphoric acid and 10 g of sulphanilamide. Dissolve completely. (Heat if necessary.) Add 0.5 g of N-1-naphthylethylenediamine dihydrochloride, and dissolve. Dilute to 1000 ml. Add 0.5 ml of Brij-35. Store in a cold, dark place. Stability: 1 month.

Extracting solution, 2 M KCl (see Operating note 1)

Potassium chloride	149.1 g
DI water	to 1000 ml

Dissolve 149.1 g of potassium chloride in about 800 ml of DI water. Dilute to 1000 ml with DI water and mix thoroughly.

Standards

Stock standard A, 100 mg N l⁻¹

Potassium nitrate	0.72 g
DI water	to 1000 ml
Chloroform	1 ml

Dissolve 0.72 g of potassium nitrate in DI water and dilute to 1000 ml. Store in a dark bottle. Add 1 ml of chloroform as a preservative.

Stock standard B, 10 mg N l⁻¹

Stock standard A	10 ml
DI water	to 1000 ml

Dilute 10 ml of stock standard A in a volumetric flask to 100 ml with DI water and mix thoroughly. Store in a dark bottle.

Working standards

Prepare working standards as required (see Operating note 1).

Operating notes

1. The standard diluent must have the same matrix as the samples and the sampler wash solution. Therefore, use extraction solution for soil analysis.

2. When soil samples in the range of 0–1 mg N l^{-1} are analysed, the dilution loop should be omitted and the 0.16 ml min^{-1} resample line is connected directly to the Sampler IV.

3. Distilled water for the dilution loops should contain 2 ml l^{-1} of Brij-35.

4. The nitrite value can be determined by eliminating the reductor column and standardizing with an appropriate nitrite value. In order to determine the nitrate values, the nitrite alone must be subtracted from the total (nitrate and nitrite).

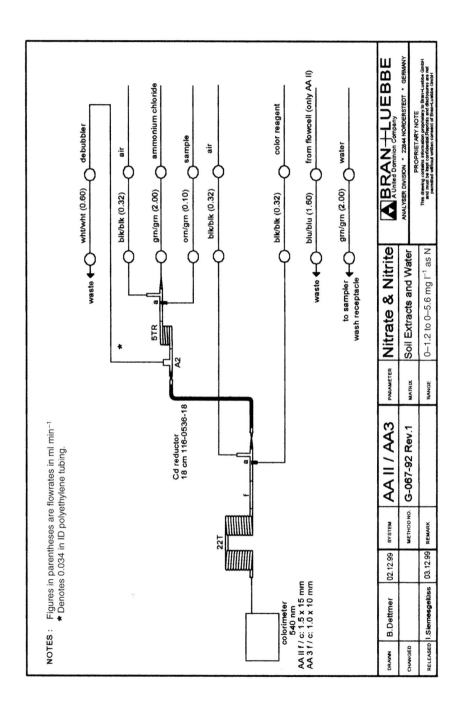

NOTES : Figures in parentheses are flowrates in ml min⁻¹
        * Denotes 0.034 in ID polyethylene tubing.

Appendix 6
Phosphate in Soil, Plant and Fertilizer Extracts[a]

Ranges: 0–1 to 0–7.5 mg l^{-1} as P and 0–6 to 0–50 mg l^{-1} as P

Description

Ortho-phosphate reacts with molybdate and ascorbic acid to form a blue compound measured at 660 nm. Antimony potassium tartrate is used as a catalyst.

A dialyser for the high range eliminates interference from coloured samples and suspended solids.

Special reagent formulations are given for Olsen soil extracts.

Hardware: 24 in Dialyser, 37°C heating bath (7.7 ml)   **Pump tubes:** 7 + 2 air + sampler wash (AAII: +1)

[a]Method No. G-103-X (© Bran+Luebbe. Not to be reproduced without permission).

Performance data

| Matrix | DI water | | NaHCO$_3$ |
Test ranges	2 mg l^{-1}	25 mg l^{-1}	10 mg l^{-1}
Sampling rate	50 h^{-1}	50 h^{-1}	60 h^{-1}
Sample: wash ratio	5:1	5:1	5:1
Sensitivity at 2 / 25 and 10 mg P l^{-1}	0.2	0.38	0.10
(sensitivity in CaCl$_2$ extracts 10% lower)			
Reagent absorbance	0.01	0.03	0.04
Coefficient of variation:			
10 replicates at 50%	0.3%	0.3%	0.3%
Pooled standard deviation			
25 at 5 levels	0.005 mg l^{-1}	0.027 mg l^{-1}	0.038 mg l^{-1}
Correlation coefficient			
(linear, 5 points)	1.000	1.000	1.000
Detection limit (water)	0.002 mg l^{-1}	0.069 mg l^{-1}	0.049 mg l^{-1}
(determined according			
to EPA procedure pt. 136, app. B)			

Note: the above performance specifications were developed with the exclusive use of genuine Bran+Luebbe parts and consumables.

Reagents

Unless otherwise stated all chemicals should be of Analytical Reagent grade or equivalent.

List of raw materials

	Safety classification
Ammonium molybdate, (NH$_4$)$_6$Mo$_7$O$_{24}$.4H$_2$O	harmful
Ammonium fluoride, NH$_4$F	toxic
Antimony potassium tartrate, K(SbO)C$_4$H$_4$O$_6$.1/2 H$_2$O	toxic
Ascorbic acid, C$_6$H$_8$O$_6$	–
Calcium chloride, CaCl$_2$.2H$_2$O	irritant
Hydrochloric acid, conc. 37%, HCl	corrosive
Potassium dihydrogen phosphate, KH$_2$PO$_4$	–
Sodium dodecyl sulphate (SDS), purest grade, C$_{12}$H$_{25}$NaO$_4$S	harmful
Sodium hydrogen carbonate, NaHCO$_3$	–
Sulphuric acid, conc. 96–98%, H$_2$SO$_4$	corrosive

Ammonium molybdate

Ammonium molybdate	1.8 g
Sulphuric acid	22.3 ml

Antimony potassium tartrate	0.05 g
Sodium dodecyl sulphate	2 g
DI water	to 1000 ml

Dissolve 1.8 g of ammonium molybdate in 700 ml of DI water. Cautiously, while swirling, slowly add 22.3 ml of sulphuric acid. Add 0.05 g antimony potassium tartrate and dilute to 1 l with DI water. Mix thoroughly and add 2 g of sodium dodecyl sulphate. Store in a dark bottle. Prepare fresh weekly.

Ascorbic acid

L-Ascorbic acid	15 g
DI water	to 1000 ml

Prepare fresh weekly. Dissolve 15 g of L-ascorbic acid in about 600 ml of DI water. Dilute to 1 l with DI water and mix thoroughly. Store in a dark bottle. Prepare fresh weekly.

Acid (see Note 1)

Sulphuric acid	22.5 ml
DI water	to 1000 ml
Sodium dodecyl sulphate	2 g

Cautiously, while swirling add 22.5 ml of sulphuric acid to about 600 ml of DI water. Cool to room temperature and dilute to 1 l. Add 2 g of sodium dodecyl sulphate and mix thoroughly. Prepare fresh weekly.

Dilution water and system wash solution

Use only sodium dodecyl sulphate (sodium lauryl sulphate) or Aerosol 22, at 2 g l^{-1}.

Sampler wash solution (see Note 5)

Use pure water without surfactant. If analysing soil extracts or other samples not dissolved in pure water use the sample matrix solution as sampler wash solution.

Extraction solution 0.01 M CaCl$_2$ (only for soil extracts, see Notes 5, 6)

Calcium chloride	1.47 g
DI water	to 1000 ml

Dissolve 1.47 g of calcium chloride dihydrate in about 600 ml of DI water. Dilute to 1 l with DI water and mix thoroughly.

Extraction solution NH₄F/HCl pH 3.5 (only for soil extracts, see Note 5)

NH₄F	1.1 g
Hydrochloric acid, conc.	2.1 ml
DI water	to 1000 ml

Dissolve 1.47 g of ammonium fluoride in about 500 ml of DI water. Add 2.1 ml of conc. hydrochloric acid and dilute to 1 l with DI water. Mix thoroughly. Adjust pH to 3.5.

Special reagents for 0.5 M NaHCO₃ soil extracts

Refer to flow diagram on page 240.

Acid

Ascorbic acid	15 g
Sulphuric acid	80 ml
DI water	1000 ml

Add carefully 80 ml of conc. sulphuric acid to about 600 ml of DI water and cool to room temperature. Dissolve 15 g of ascorbic acid and dilute to 1 l with DI water and mix thoroughly. Store in a dark bottle in the refrigerator. The solution is stable for 1 week.

Ammonium molybdate

Ammonium molybdate	1.8 g
Antimony potassium tartrate	0.05 g
DI water	to 1000 ml
Sodium lauryl sulphate	2 g

Dissolve 1.8 g of ammonium molybdate and 0.05 g of antimony potassium tartrate in about 800 ml of DI water. Dilute to 1 l with DI water and add 2 g of sodium lauryl sulphate. Mix thoroughly. Store in a dark bottle. The solution is stable for a month before the SDS is added. The solution must be colourless. The ammonium molybdate must be perfectly white, with no green tint. Ultra-pure SDS is critical to good method performance.

Extraction solution 0.5 M NaHCO₃ (pH 8.5) (see Note 5)

Sodium hydrogen carbonate	42 g
Sodium hydroxide	as required
DI water	to 1000 ml

Dissolve 42 g of sodium hydrogen carbonate in about 500 ml of DI water. Adjust pH to 8.5 using sodium hydroxide. Dilute to 1 l with DI water.

Standards

Stock phosphate standard, 1000 mg l⁻¹ as P

Potassium dihydrogen phosphate	4.394 g
DI water	to 1000 ml

Dissolve 4.394 g of potassium dihydrogen phosphate in about 200 ml of DI water. Dilute to 1 l with DI water.

Prepare working standards as required. If analysing soil extracts or other samples not dissolved in pure water, dilute the working standards with the sample matrix solution.

Notes

1. Wash the manifold daily as follows:
- (a) If using a stainless steel probe, remove it and place it in a 10% solution of nitric acid.
- (b) Pump wash solution through all reagent lines for at least 5 min.
- (c) Pump sodium hypochlorite solution (up to 10% available chlorine) through the sample line for 5 min, while pumping wash solution through the other lines.
- (d) Replace the sample probe.
- (e) Pump water through the sample line for 10 min, while continuing to pump wash solution through the reagent lines.

2. Use only the specified surfactants, as excess drift and carryover may otherwise result.

In particular, Brij-35 reacts with molybdate, causing precipitation and drift, while Triton X-100 significantly suppresses the reaction.

3. The standard diluent and the sampler wash solution must have the same matrix as the samples. Therefore, if analysing soil samples use the soil extraction solution (0.01 M $CaCl_2$, NH_4F/HCl or 0.5 M $NaHCO_3$) as diluent for the standards and as sampler wash solution.

4. The sensitivity in $CaCl_2$ extracts is 10% lower.

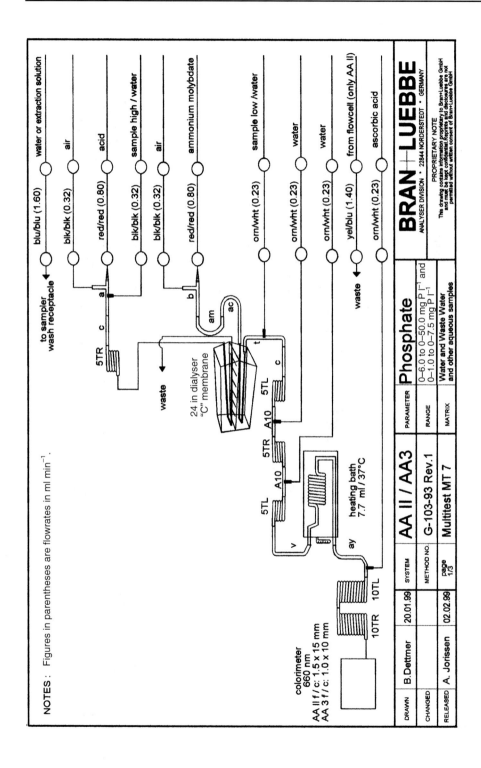

NOTES : Figures in parentheses are flowrates in ml min⁻¹ .

NOTES : Figures in parentheses are flowrates in ml min^{-1}.

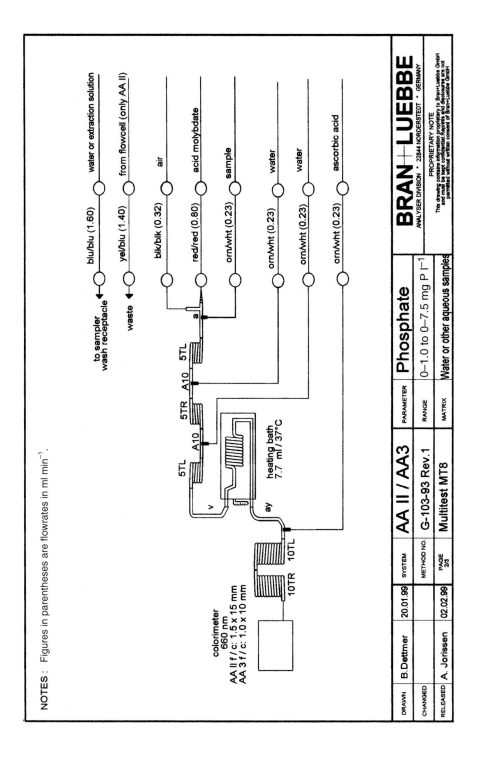

DRAWN	B.Dettmer	20.01.99	SYSTEM	AA II / AA3	PARAMETER	Phosphate
CHANGED			METHOD NO.	G-103-93 Rev.1	RANGE	0–1.0 to 0–7.5 mg P l^{-1}
RELEASED	A. Jorissen	02.02.99	PAGE 2/3	Multitest MT8	MATRIX	Water or other aqueous samples

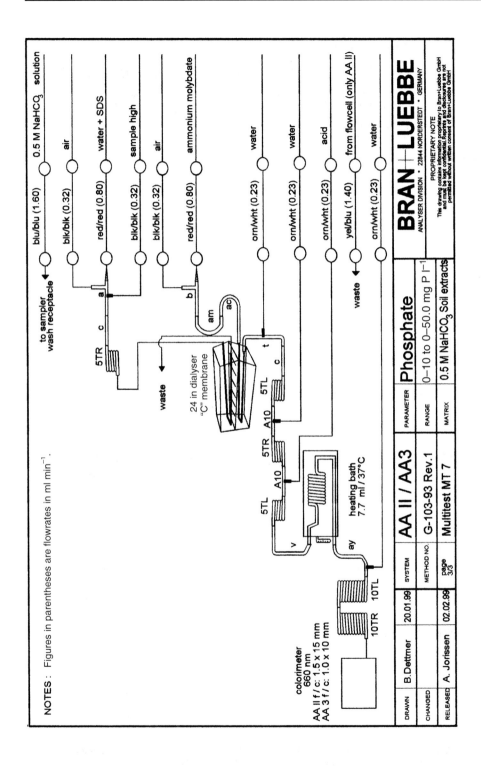

Appendix 7
Analytical Methods Used by ADAS for the Analysis of Organic Manures[a]

Sample preparation and storage

Homogenization of the whole sample before sub-sampling. Original sample and sub-samples are stored refrigerated. Maceration and homogenization of sub-samples before use. This is potentially the stage when most errors can occur. Samples containing both fluid and fibrous fractions may be very difficult to homogenize.

Ammonium-N

Water extraction or dilution of fresh material. Alkaline steam distillation with final measurement by titration.

Dry matter/total solids

Oven drying; difference in weight between ambient temperature and 102°C.

---

[a]R. Llewelyn and S. Richardson (Wolverhampton, 2001, personal communication).
(Reproduced with permission of ADAS, Wolverhampton, 17 April 2001.)

Nitrate-N

Water extraction or dilution of fresh material. Alkaline steam distillation with Devarda's alloy after removal of ammonium N. Final measurement by titration.

pH

Water suspension or direct. pH electrode and meter at 20°C.

Total N

Kjeldahl digestion of fresh material (with addition of a strong reducing agent if high nitrate is expected). Alkaline steam distillation with final measurement by titration.

Total P, K, Mg and S

Digestion with boiling *aqua regia*. Dilution and centrifugation with final measurement by ICP-OES (inductively coupled plasma optical emission spectroscopy).

Uric acid-N

Buffer solution extraction of fresh material with final measurement by HPLC.

The above methods are designed to measure the total concentrations of each parameter. Although the various fractions of readily available nitrogen are extracted with water or buffer solutions, the intention is to measure the total concentration of each parameter as described. Other extraction methods and measurement techniques may be equally effective and acceptable if they produce comparable values. NIRS, in particular, has been demonstrated to offer considerable potential for very rapid analysis of DM, total N, ammonium-N and organic matter.

Harmonized Reporting of Analysis Results for Organic Manures

Following a seminar at ADAS Wolverhampton on 28 March 2001 which endeavoured to established a consistent format for results from analytical laboratories, the following points were established:

Clarity

The first priority is to keep the report simple and clear to enable the farmer to understand and use the results.

Analytes

- *Essential* – dry matter, total N, P, K and ammonium-N; plus uric acid-N for poultry manures.
- *Desirable* – total S, Mg, and pH. Nitrate-N for old manures stacked for more than 3 or 4 months and composted organic wastes.

Fresh weight

Report results on a fresh weight basis, since that is how farmers apply organic manures to land. Mass/mass for solid manures, mass/volume for slurries, with the cut-off point at approx. 10% DM content.

Oxides

N and fractions of readily available N are reported as the element. P, K, S and Mg are reported as the oxide, since this is a statutory requirement for inorganic fertilizers, and fertilizer recommendations for the latter four nutrients are given in terms of the oxide content in the literature.

Laboratory results

The two columns of laboratory results should be separated from the three columns of fertilizer values, either side-by-side or one over the other. To keep the report simple, laboratories may prefer not to present the laboratory results since the farmer is normally only interested in the fertilizer values. Some quality management or accreditation systems may require reporting of the laboratory results. These could be reported as the element, as shown here, or as the oxide for P, K, S and Mg.

Fertilizer values

This is usually the only part of the report the farmer is interested in. Thus columns 3 and 4 in the combined table are the most important part of the report. These could be highlighted to add clarity.

Metric units

See column 4 below. Report fertilizer values as kg t^{-1} (solids) or kg m^3 (liquids).

Imperial units

See column 5 below. It is desirable to also include results in imperial units because many farmers, and possibly agricultural consultants, continue to think of manure use in imperial terms. Thus, 1 unit = 1% of one hundredweight = 1.12 lb (pounds), or 0.488 kg.

Calculation of Fertilizer Values of Organic Manure from Laboratory Analysis Data

Laboratory results may be reported on an 'as received' by volume basis or 'dry matter' by mass basis, depending on whether the sample is considered to be a solid or a liquid. Samples >10% DM are generally treated as solids, and <10% DM as liquids. However, there are frequently exceptions to this rule, e.g. some fibrous samples <10% DM are difficult to pipette, so are analysed on a weight basis, but reported on a volume basis. Some customers can pump thick samples with >10% DM, so need the fertilizer values reported as liquids on a volume basis. Routine determination of density values facilitates calculation and reporting of fertilizer values in either mass or volume units.

Units for Reporting Analysis Data for Organic Manures

Solid samples

Parameter	Lab unit	Multiply by	Fert. value (metric)
Dry matter	g kg^{-1}	0.1	% Dry matter
Total N	g kg^{-1}	% DM $\times$ 0.01	kg t^{-1} N
Ammonium-N	µg kg^{-1}	% DM $\times$ 0.00001	kg t^{-1} N
Nitrate-N	µg kg^{-1}	% DM $\times$ 0.00001	kg t^{-1} N
Uric-N	µg kg^{-1}	% DM $\times$ 0.00001	kg t^{-1} N
Phosphorus (P)	g kg^{-1}	% DM $\times$ 0.0229	kg t^{-1} P$_2$O$_5$
Potassium (K)	g kg^{-1}	% DM $\times$ 0.01205	kg t^{-1} K$_2$O
Magnesium (Mg)	g kg^{-1}	% DM $\times$ 0.01658	kg t^{-1} MgO
Sulphur (S)	g kg^{-1}	% DM $\times$ 0.025	kg t^{-1} SO$_3$

Note: for imperial units, kg t^{-1} $\times$ 1.9988 = units/ton.

Liquid samples

Parameter	Lab unit	Multiply by	Fert. value (metric)
Total solids	g l^{-1}	0.1	% Dry matter
Total N	g l^{-1}	1.0	kg m^{-3} N
Ammonium-N	µg l^{-1}	0.001	kg m^{-3} N
Nitrate-N	µg l^{-1}	0.001	kg m^{-3} N
Uric-N	µg l^{-1}	0.001	kg m^{-3} N
Phosphorus (P)	g l^{-1}	0.00229	kg m^{-3} P$_2$O$_5$
Potassium (K)	g l^{-1}	0.001205	kg m^{-3} K$_2$O
Magnesium (Mg)	g l^{-1}	0.001658	kg m^{-3} MgO
Sulphur (S)	g l^{-1}	0.0025	kg m^{-3} SO$_3$

Note: for imperial units, kg m^{-3} × 8.942 = units/1000 gallons.

Typical Reports

Typical report on fertilizer value of organic manure
[Solid sample: FYM, sludge, compost, industrial waste – reported by weight]

Laboratory results (mass DM basis)			Fertilizer values (mass/mass fresh basis)	
			Metric	Imperial
pH	7.5	pH	7.5	7.5
Dry matter	250 g kg^{-1}	Dry matter	25% (m/m)	25% (m/m)
Ammonium-N	155 mg kg^{-1}			
Nitrate-N	15 mg kg^{-1}			
Uric acid-N	100 mg kg^{-1}			
		Readily available N (as N)	0.07 kg t^{-1}	0.14 units/ton
Total N (as N)	13 g kg^{-1}	Total nitrogen (as N)	3.25 kg t^{-1}	6.50 units/ton
Total P (as P)	19.2 g kg^{-1}	Phosphate (as P$_2$O$_5$)	10.99 kg t^{-1}	21.98 units/ton
Total K (as K)	12.4 g kg^{-1}	Potash (as K$_2$O)	3.74 kg t^{-1}	7.48 units/ton
Total Mg (as Mg)	1.4 g kg^{-1}	Magnesium (as MgO)	0.58 kg t^{-1}	1.16 units/ton
Total S (as S)	2.2 g kg^{-1}	Sulphur (as SO$_3$)	1.38 kg t^{-1}	2.76 units/ton

FYM = farmyard manure.

Typical report on fertilizer value of organic manure
[Liquid sample: slurry, effluent, dirty water – reported by volume]

Laboratory results (mass/volume fresh basis)		Fertilizer values (mass/volume fresh basis)		
			Metric	Imperial
pH	6.0	pH	6.0	6.0
Total solids	10 g l⁻¹	Dry matter	1.0% (m/v)	1.0% (m/v)
Ammonium-N	257 mg l⁻¹			
Nitrate-N	15 mg l⁻¹			
Uric acid-N	40 mg l⁻¹			
		Readily available N (as N)	0.31 kg m⁻³	2.78 units/1000 gal
Total N (as N)	0.40 g l⁻¹	Total nitrogen (as N)	0.40 kg m⁻³	3.56 units/1000 gal
Total P (as P)	130 mg l⁻¹	Phosphate (as P_2O_5)	0.30 kg m⁻³	2.65 units/1000 gal
Total K (as K)	412 mg gl⁻¹	Potash (as K_2O)	0.50 kg m⁻³	4.42 units/1000 gal
Total Mg (as Mg)	500 mg l⁻¹	Magnesium (as MgO)	0.83 kg m⁻³	7.41 units/1000 gal
Total S (as S)	300 mg l⁻¹	Sulphur (as SO_3)	0.75 kg m⁻³	6.68 units/1000 gal

Appendix 8
Laboratory Safety

Health and safety in a laboratory environment is the moral responsibility of all, but is also a legal responsibility for employers and supervisors. Space prohibits more than a few guidelines, and indeed, many books have been written on the subject. These include:

- Luxon, S.G. (ed.) (1992) *Hazards in the Chemical Laboratory*, 5th edn. Royal Society of Chemistry, Cambridge, 676 pp.
    Details at: http://www.rsc.org/is/books/books1.htm
- Committee on Chemical Safety (1995) *Safety in Academic Chemistry Laboratories*, 6th edn. American Chemical Society, Washington, DC, 70 pp.
- Furr, A.K. (ed.) (2000) *CRC Handbook of Laboratory Safety*, 5th edn. CRC Press, Boca Raton, Florida, 808 pp.
    Details at: http://www.crcpress.com/
- Mercier, P. (ed.) (1996) *Laboratory Safety Pocket Handbook*, Genium Publishing Corporation, Amsterdam, New York.
    Details at: http://www.genium.com/

Organizations should be able to show that a safety management system is in operation. This will involve setting up departmental safety committees with representatives from various areas, e.g. senior staff, technical staff, administrative staff, portering staff and postgraduate students. The committee should arrange regular safety inspections, make recommendations based on these and any other reported incidents, and monitor progress on remedial action. Minutes should be rapidly circulated and a summary put on notice boards and the intranet (as well as e-mails to appropriate staff) in order to inform personnel as quickly as possible of the latest safety recommendations. The following advisory booklet is available:

> *Health and Safety Management in Higher and Further Education:*
> *Guidance on Inspection, Monitoring and Auditing.* (1992), HSE
> Education Services Advisory Committee, HMSO, London, 16 pp.,
> ISBN 011 886315 0 (Price £3.00).

Implications of the Workplace (Health, Safety and Welfare) Regulations 1992 necessitate one accessing the safety guidance for the education sector, with a list of publications, which is available at:

> http://www.hse.gov.uk/pubns/iacl97.htm#2

In the US, the health, safety and environmental affairs (HS&E) policy is administered by the Occupational Safety and Health Administration, whose website is:

> http://www.osha.gov/index.html

Of importance is the OSHA Laboratory Standard, 29 CFR 1910.1450, which is similar to the UK COSHH regulations (see below), and deals with occupational exposure to hazardous chemicals in laboratories. Information on various aspects of this Laboratory Standard is found at:

> http://www.osha-slc.gov/SLTC/laboratories/index.html

and the topic of personal protective equipment is found at:

> http://www.osha-
> slc.gov/SLTC/personalprotectiveequipment/index.html

The controversial question of wearing contact lenses in the laboratory is discussed at:

> http://pubs.acs.org/hotartcl/chas/97/mayjun/con.html

OSHA believes that contact lenses do NOT pose additional hazards to the wearer, but additional appropriate protection, such as in the form of goggles or a visor is necessary.

A list of educational safety books is listed by the Laboratory Safety Institute (USA) at:

> http://www.labsafety.org/pubprodorder.htm

They also give 40 steps to laboratory safety at the website:

> http://www.labsafety.org/40steps.htm

In the UK, there are two main components to laboratory safety management: risk assessment, and Control of Substances Hazardous to Health (COSHH). The latter aspect is a legal requirement of The Control of Substances Hazardous to Health Regulations 1999 (COSHH).

Risk Assessment

Risk assessment applies to all activities in the laboratory, including lifting heavy equipment and use of furnaces, for example, whether or not any hazardous substances are involved. The five basic steps in risk assessment are:
1. Look for the hazards.

2. Determine who might be harmed and how.
3. Evaluate the risks and decide whether existing precautions are adequate or whether further safeguards are necessary.
4. Record your findings.
5. Regularly review the assessment and revise if necessary.

COSHH

Full details are given in the HSE information leaflets obtainable as given below; see also RSC (1996). Before the commencement of any work that involves or creates substances that may be hazardous to health, the risks, control measures and disposal procedures to be adopted should be assessed and recorded on appropriate forms. These should be signed by the assessor, and where appropriate by the student, and then countersigned by the Head of Department. One copy should be kept for local reference in the laboratory, with the original filed centrally for inspection by external inspectors. Only after this, should any new chemicals be ordered by a designated person (or deputy) who should check that the COSHH form has been duly completed. COSHH guidelines may be summarized in seven points:

1. Assess risks.
2. Decide precautions.
3. Control exposure.
4. Ensure use of control measures.
5. Monitor exposure.
6. Carry out appropriate health surveillance.
7. Ensure the training, informing and supervision of employees.

Electrical Testing

All equipment should be tested with a portable appliance tester at regular intervals, and an attached label should give the date of the test and whether the result was Pass or Fail; bar-coded testing systems are available. Static equipment may be fully tested every 2–4 years, but a recorded visual check should be made more frequently, and ideally each time the item is used. Care should be taken not to apply high test voltages to sensitive equipment (e.g. computers), and some testing devices have special settings to be used for this purpose. Equipment that is often moved to different locations should be fully tested much more frequently. In the UK, one must comply with the Electricity at Work Regulations 1989. Even if the keeping of records is not a legal requirement, it would be of great benefit in the event of an accident, with subsequent litigation, to be able to prove that everything reasonable was done to ensure the safety of personnel. See the booklet:

Maintaining Portable Electrical Equipment in Offices and other Low-risk Environments. (1999), HSE, Sudbury, 12 pp.
HSE references to electrical safety are found at:
http://www.hse.gov.uk/pubns/elecindx.htm, and the booklet is downloadable from:
http://www.hse.gov.uk/pubns/indg236.pdf

The Health and Safety Executive

Matters of health and safety in the workplace are administered by the Health and Safety Executive (HSE). They supply safety publications either free or purchasable. These are available from:

HSE Books, PO Box 1999, Sudbury, Suffolk CO10 6FS, UK. Tel. +44 (0) 1787 881165; Fax +44 (0) 1787 313995

Information leaflets on risk assessment may also be accessed from:
http://www.hse.gov.uk/signpost/content/r.htm
whereas leaflets about COSHH are accessible from:
http://www.hse.gov.uk/signpost/content/c.htm
Other topics may be searched alphabetically from:
http://www.hse.gov.uk/signpost/index.htm
A Guide to Risk Assessment Requirements (August, 2001), and *Five Steps to Risk Assessment* (May, 1998) may be downloaded as .pdf files (http://www.hse.gov.uk/pubns/indg218.pdf and http://www.hse.gov.uk/pubns/indg163.pdf) from: http://www.hse.gov.uk/pubns/raindex.htm
COSHH – a Brief Guide to the Regulations is similarly downloadable at:
http://www.hse.gov.uk/pubns/indg136.pdf

Other regulations which may be applicable are:

- *Management of Health and Safety at Work Regulations 1999* (Management Regulations)
- *Manual Handling Operations Regulations 1992* (Manual Handling Regulations)
- *Personal Protective Equipment at Work Regulations 1992* (PPE)
- *Health and Safety (Display Screen Equipment) Regulations 1992* (Display Screen Regulations)
- *Control of Asbestos at Work Regulations 1987* (Asbestos Regulations)
- *Control of Lead at Work Regulations 1998* (Lead Regulations)
- *Electricity at Work Regulations 1989*
- *Workplace (Health, Safety and Welfare) Regulations 1992*

Other regulations apply to radioactive chemicals, which are outside the scope of this volume.

Appendix 9
Chemical Composition Data Sources for Plants, Feeds, Blood, Urine and Soils

It is sometimes helpful when designing methods and making up standards to have an idea as to the possible range of concentrations of elements or compounds in a type of sample that is new to the laboratory. Some sources of information are described.

Plant Constituents

A remarkable old German publication in two volumes (Wehmer, 1929, 1931) lists hundreds of plant species with many of their chemical constituents, including ash composition, tannins, sugars, alkaloids, etc. Where figures are omitted, copious references are given.

Of UK interest are two older publications with data on the mineral composition of grassland herbage. The first was published by the Agricultural Research Council, and was essentially a comparison of analytical methods for mineral elements in various leaves and pasture herbage (ARC, 1963); the second was from the Grassland Research Institute, Hurley, and the Welsh Plant Breeding Station, Aberystwyth (now merged as IGER, Aberystwyth) (Whitehead, 1966).

A recent publication by James A. Duke lists the phytochemical constituents of generally-regarded-as-safe (GRAS) herbs (Duke, 2000a). It is also available as a searchable database on disk (WordPerfect™ 5.1 macros). There is also a volume dealing with 365 herbs having medicinal or folklore medicinal properties (Duke, 2000b).

Feed Constituents

The proximate analyses, mineral composition, digestibility coefficients and nutritive values of brassicas, by-products, grains, grasses, hays, legumes, oil cakes, root crops, seeds and silages are given in MAFF/ADAS (1971) *Nutrient Allowances and Composition of Feedingstuffs for Ruminants* (Advisory Paper No. 11).

A more recent analytical tabulation covering individual trace elements, amino acids, and volatile fatty acids, together with proximate analyses, ADF, MADF, NDF, cellulose, lignin, starch, water soluble carbohydrates, etc., has the title *UK Tables of Nutritive Value and Chemical Composition of Feedingstuffs* (MAFF, 1990).

Proximate analyses, ADF, NDF, minerals and energy values for many cattle feeds are listed by Perry (1980).

By-products

The proximate analyses of many types of by-products used in livestock feed (e.g. brewers' grains, coffee grounds, olive cake, peanut hulls, wood pulp) together with digestibility and energy values are given by Boer and Bickel (1988).

Blood Analytes

The chemical constituents of the blood of large animals is tabulated in the volume edited by Kaneko (1989) *Clinical Biochemistry of Domestic Animals.*

Urine Analytes

The normal concentrations of urine constituents in domestic animals (cat, cow, dog, goat, horse, pig and sheep) compiled by J.J. Kaneko are given by Stahr (1991) in *Analytical Methods in Toxicology.*

Trace Elements in Soils and Plants

A comprehensive treatment of the properties and contents of trace elements in soils and plants is given by Kabata-Pendias (2000).

Appendix 10
Atomic Weights, Units and Conversion Tables

Table of Atomic Weights[a]

Name	Atomic number	Symbol	Atomic weight	Valency
Aluminium	13	Al	26.982	3
Antimony	51	Sb	121.76	3,5
Arsenic	33	As	74.922	3,5
Barium	56	Ba	137.33	2
Bismuth	83	Bi	208.98	3,5
Boron	5	B	10.811	3
Bromine	35	Br	79.904	1,3,5,7[b]
Cadmium	48	Cd	112.41	2
Calcium	20	Ca	40.078	2
Carbon	6	C	12.011	2,4
Caesium	55	Cs	132.91	3,4
Chlorine	17	Cl	35.453	1,3,5,7[b]
Chromium	24	Cr	51.996	2,3,6
Cobalt	27	Co	58.933	2,3
Copper	29	Cu	63.546	1,2
Fluorine	9	F	18.998	1
Gold	79	Au	196.97	1,3
Hydrogen	1	H	1.0079	1
Iodine	53	I	126.90	1,3,5,7[b]
Iron	26	Fe	55.845	2,3
Lanthanum	57	La	138.91	3
Lead	82	Pb	207.2	2,4

Name	Atomic number	Symbol	Atomic weight	Valency
Lithium	3	Li	6.941	1
Magnesium	12	Mg	24.305	2
Manganese	25	Mn	54.938	2,3,4,6,7
Mercury	80	Hg	200.59	1,2
Molybdenum	42	Mo	95.94	3,4,6
Nickel	28	Ni	58.69	2,3
Nitrogen	7	N	14.007	3,5
Oxygen	8	O	15.999	2
Phosphorus	15	P	30.974	3,5
Potassium	19	K	39.098	1
Selenium	34	Se	78.96	2,4,6
Silicon	14	Si	28.086	4
Silver	47	Ag	107.87	1
Sodium	11	Na	22.990	1
Strontium	38	Sr	87.62	2
Sulphur	16	S	32.066	2,4,6
Tin	50	Sn	118.71	2,4
Titanium	22	Ti	47.87	3,4
Vanadium	23	V	50.942	3,5
Zinc	30	Zn	65.39	2

[a]Atomic weights (relative atomic masses, A_r) adapted from IUPAC, *Pure and Applied Chemistry* 68(12), 2339 (1996), based on $^{12}C = 12$, and given to 5 significant figures.

[b]Halogen (X) valency states of 3, 5 and 7 only exist in XF_3, XF_5, XF_7 and oxo compounds.

Conversion Table for SI Units

(see http://physics.nist.gov/cuu/Units/units.html; and http://www.bipm.fr/enus/welcome.html
http://www.bipm.fr/pdf/si-brochure.pdf)

Non-SI units		SI units		Conversion factor
Name	Symbol	Preferred unit	Symbol	
Length:				
yard	yd	metre	m	1 yd = 0.914 m
foot	ft	metre	m	1 ft = 0.305 m
inch	in	millimetre	mm	1 in = 25.4 mm
centimetre	cm	millimetre	mm	1 cm = 10 mm
micron	μ	micrometre	μm	1 μ = 1 μm = 10^{-6} m
millimicron	mμ	nanometre	nm	1 mμ = 1 nm = 10^{-9} m
Ångström	Å	nanometre	nm	10 Å = 1 nm
Area:				
hectare	ha	square metre	m^2	1 ha = 10^4 m^2
acre		square metre	m^2	1 acre = 0.405 ha

Non-SI units		SI units		Conversion factor
Name	Symbol	Preferred unit	Symbol	
square yard	sq. yd	square metre	m^2	1 sq. yd = 0.836 m^2
square foot	sq. ft	square metre	m^2	1 sq. ft = 0.093 m^2
square inch	sq. in	square millimetre	mm^2	1 sq. in = 645.2 mm^2
Mass:				
ton (avoirdupois.)	t	kilogram	kg	1 t = 1016 kg
tonne (metric)	t	kilogram	kg	1 tonne = 10^3 kg
hundredweight	cwt	kilogram	kg	1 cwt = 50.8 kg
pound	lb	kilogram	kg	1 lb = 0.4536 kg
ounce	oz	gram	g	1 oz = 28.35 g
Volume:				
gallon (UK)		cubic decimetre	dm^3	1 gallon = 4.546 dm^3
pint (UK)		cubic decimetre	dm^3	1 pint = 0.568 dm^3
litre	l	cubic decimetre	dm^3	1 l = 1 dm^3 = 10^{-3} m^3
millilitre	ml	cubic centimetre	cm^3	1 ml = 1 cm^3 = 10^{-6} m^3
microlitre	μl	cubic millimetre	mm^3	l μl = 1 mm^3 = 10^{-9} m^3
Quantity:density				
percent (w/v) or (m/v); (m/V)	% (w/v)	grams per cubic decimetre	g dm^{-3}	1%(w/v) = 10 g dm^{-3}
grams per litre	g l^{-1}	grams per cubic decimetre	g dm^{-3}	1 g l^{-1} = 1 g dm^{-3}
milligrams per litre	mg l^{-1}	milligrams per cubic decimetre	mg dm^{-3}	1 mg l^{-1} = 1 mg dm^{-3}
parts per million (w/v)	ppm	milligrams per cubic decimetre	mg dm^{-3}	1 ppm = 1 mg dm^{-3}
parts per million (w/w)	ppm	kilograms per kilogram	kg kg^{-1}	1 ppm = 1 mg kg^{-1}
Quantity: pressure				
bar	bar	megapascal	MPa	1 bar = 0.1 MPa
millimetre mercury	mmHg	pascal	Pa	1 mmHg = 133 Pa
Quantity:amount of matter				
gram atom	grat	mol	mol	1 grat = 1 mol atoms
gram molecule	gmol	mol	mol	1 gmol = 1 mol molecules
equivalent	Eq	mol	mol	1 Eq = 1 mol monovalent ions
Quantity: concentration				
mol per litre (molarity)	mol^{-1}	mol per cubic decimetre	mol dm^{-3}	1 mol/l = 1 M = 1 mol dm^{-3}
normality	Eq^{-1} (N)	mol per cubic decimetre	mol dm^{-3}	1 N = 1 mol dm^{-3} of monovalent ions
mol per cent	mol %	mol per cubic decimetre	mol dm^{-3}	1 mol % = 10 mol dm^{-3}

Non-SI units		SI units		Conversion factor
Name	Symbol	Preferred unit	Symbol	
Quantity:matter content				
milliequiv. per 100 g	mEq/ 100 g	centimol per kilogram	cmol kg^{-1}	1 mEq/100 g = 1 cmol kg^{-1}
mol per cent	mol %	mol per kilogram	mol kg^{-1}	1 mol% = 10 mol kg^{-1}
molality	molal (m)	mol per kilogram	mol kg^{-1}	1 molal = 1 mol kg^{-1}

Prefixes used with SI units

Name	Symbol	Meaning	Name	Symbol	Meaning
exa	E	10^{18}	deci[a]	d	10^{-1}
peta	P	10^{15}	centi[a]	c	10^{-2}
tera	T	10^{12}	milli	m	10^{-3}
giga	G	10^{9}	micro	μ	10^{-6}
mega	M	10^{6}	nano	n	10^{-9}
kilo	k	10^{3}	pico	p	10^{-12}
hecto[a]	h	10^{2}	femto	f	10^{-15}
deca[a]	da	10^{1}	atto	a	10^{-18}

[a]These prefixes are permitted but their use is discouraged.

Other Approximate Conversion Factors

pounds per acre × 1.12	=	kilograms per hectare
kilograms per hectare × 0.89	=	pounds per acre
10 tonnes per hectare	=	4 tons per acre
100 kg per hectare	=	80 units per acre
1 unit per acre	=	1.25 kg per hectare
1 fertilizer Unit	=	1.12 lb = 0.51 kg
1 hectare = 2.471 acres	=	1.076×10^5 sq. ft
1 cubic metre	=	220 gallons (UK)
1 cubic metre per hectare	=	90 gallons per acre
gallons per acre × 11.233	=	litres per hectare
1 fluid ounce = 0.05 pint	=	0.02841 litres = 28.41 ml
P_2O_5 × 0.4364	=	P
P × 2.2915	=	P_2O_5
K_2O × 0.8301	=	K
K × 1.2047	=	K_2O
CaO × 0.7146	=	Ca
Ca × 1.3994	=	CaO
MgO × 0.6031	=	Mg
Mg × 1.6581	=	MgO
Irish acres × 0.656	=	hectares

Sintered Glass Porosity Tables

Comparative porosities.

Porosity (former BS 1752 Grade)	00	0	1	2	3	4	5
ISO 4793 Designation	P500	P250	P160	P100	P40	P16	P1.6
ISO 4793 Pore diameter (µm)	250–500	160–250	100–160	40–100	16–40	10–16	1.0–1.6

ASTM/BS porosities.

Description ASTM/BS	Pore size (µm)
Extra coarse (EC)	170–220
Coarse (C)	40–60
Medium (M)	10–16
Fine (F) BS	4–10
Fine (F) ASTM	4–5.5
Very Fine (VF)	2–2.5
Ultra Fine (UF)	0.9–1.4

Subject Index

Commercial Index